Urban and Landscape Perspectives

Volume 19

Aims and Scope

Urban and Landscape Perspectives is a series which aims at nurturing theoretic reflection on the city and the territory and working out and applying methods and techniques for improving our physical and social landscapes.

The main issue in the series is developed around the projectual dimension, with the objective of visualising both the city and the territory from a particular viewpoint, which singles out the territorial dimension as the city's space of communication and negotiation.

The series will face emerging problems that characterise the dynamics of city development, like the new, fresh relations between urban societies and physical space, the right to the city, urban equity, the project for the physical city as a means to reveal civitas, signs of new social cohesiveness, the sense of contemporary public space and the sustainability of urban development.

Concerned with advancing theories on the city, the series resolves to welcome articles that feature a pluralism of disciplinary contributions studying formal and informal practices on the project for the city and seeking conceptual and operative categories capable of understanding and facing the problems inherent in the profound transformations of contemporary urban landscapes.

More information about this series at http://www.springer.com/series/7906

Grazia Concilio • Francesca Rizzo
Editors

Human Smart Cities

Rethinking the Interplay between Design and Planning

Springer

Editors
Grazia Concilio
Department of Architecture and Urban Studies
Politecnico di Milano
Milan, Italy

Francesca Rizzo
Department of Architecture and Territorial Development
University of Bologna
Bologna, Italy

Urban and Landscape Perspectives
ISBN 978-3-319-33022-8   ISBN 978-3-319-33024-2 (eBook)
DOI 10.1007/978-3-319-33024-2

Library of Congress Control Number: 2016945266

Cover Image: Passages (n°63), CCCB, Barcelona (2007) by Roberto Manuelli.

Printed on acid-free paper

This Springer imprint is published by Springer Nature
The registered company is Springer International Publishing AG Switzerland

Preface: Experimenting Between Design and Planning

Urban transformation is widely recognized as a complex phenomenon, rich in uncertainty. It is the unpredictable consequence of the complex interplay involving urban forces (both top-down or bottom-up), urban resources (spatial, social, economic, and infrastructural, as well as political or cognitive), and transformation opportunities (endogenous or exogenous).

The recent attention to urban living lab and human smart city initiatives has revealed a promising bridge between urban governance mechanisms and the microscale environments and dynamics of such forces and resources. This bridge is represented by urban collaborative environments, in which processes of smart service co-design take place through dialogic interaction with and among citizens within a situated and cultural-specific frame.

As a response to new emerging needs and ways of generating value, over the last decades, the discipline of design – traditionally bound to the development of tangible artifacts – has expanded its focus to include intangible artifacts such as signs, interactions, processes, and services. This evolution entailed the generation of a wide set of tools and methods, primarily meant to help designers give shape to intangible outcomes (such as processes and interactions) and to their tangible substrates (usually called "touch points" in the service sciences). In quite a few cases, shaping these new intangible outcomes requires the contextual development of a network of actors who will contribute to their realization. The need to align, with a single vision, actors and stakeholders who may have different and sometimes conflicting needs and goals characterizes design practice in some of its most recent developments. Complex participatory methods have been developed and experimented in the field of service design, giving birth to a wide set of tools for the involvement of actors, the construction of networks, the definition of the underpinned business models, and the prototyping and testing of services.

This evolution led to a considerable expansion of the interest fields of the design discipline, introducing design methods and design thinking into different contexts and starting a productive dialogue with other disciplines.

The most interesting among these is the relation with urban planning, which has a long tradition of dialogue with actors and stakeholders, primarily based on the idea of aligning municipalities and citizens with regard to the strategic decisions linked to the transformation of the urban environment. The massive introduction of digital services, which generated a new intangible city layer, was the natural meeting place for design, urban planning, and smart services. The concept of "smart city" boosted the relation between the two disciplines. It also introduced the idea that the harmonious development of contemporary cities must be based on the capability to design and manage the interaction between their traditional physical structure and the new digital information infrastructure. This is to be achieved by introducing "smart" service ideas and solutions for the citizens' emerging needs. These new urban services reverse the relationship between the material substrate and the digital layer of the city: they do not just fit into the existing spaces, but actually modify the physical substrate and remodel the city by changing the ways in which people produce it.

Within the most recent discussion on smart cities and the way this vision is affecting urban changes and dynamics, this book explores the interplay between planning and design at the level of the theories and practices of their domains.

The book title is deeply related to the conceptual output of two European projects[1] funded by the Competitive and Innovation Framework Programme (CIP), where the bridge between the two disciplines represented the important methodological and conceptual challenge in the development of smart city policies and services. Therefore the book collects contributions from a group of thinkers and practitioners who in different ways (participation in projects, conferences, and workshops) contributed to the conceptualization of the human dimension of urban smartness.

The book is divided into four parts. The first one, entitled "Needs for New Durable Visions," confirms the strategic importance of visions for cities. When durable and rooted visions are available, the policies themselves become occasions for experiments, allowing new modes of city-making to be envisioned, verified, and implemented. The three contributions in this part discuss the importance of strategic visions, at the same time introducing new ones that increasingly show their potentials and importance in urban contexts.

The chapter by **Louis Albrechts** argues that new conceptual infrastructures and transformative practices are needed to cope with contemporary challenges. It reflects first on the nature and characteristics of transformative practices that construct images/visions of a preferred innovative structural outcome and opportunities for

---

[1]Peripheria European Project (2010–2013; Grant Agreement No. 271015) which aimed at involving specific competences in urban planning and design for the conception of new people-centered services that would also represent "smartness" as the capability of cities to develop solutions in line with citizen needs and desires.

My Neighbourhood (2013–2015, Grant Agreement No.: 325227) is an EUfunded research project which started in January 2013. Its goal was to apply service design methods and tools in 4 different European neighborhoods (Lisbon, Milan, Aalborg and Birmingham) to identify and support the establishment and the upscale of grassroots and community-based initiatives, through the adoption of a web-based service platform.

their implementation. It then deals with the political-economic context and proposes envisioning as a learning process, discussing (among other things) transformative triggers and the power of visions. This is followed by a brief description of two cases, Hasselt and Antwerp, which are relevant in terms of transformative actions and visions. Finally, it provides ingredients for more radical strategic planning as the governance of collective affairs, which requires planners to cultivate activist modes of planning and to be more than navigators keeping the ship on course.

Grazia Concilio explores some of the most significant potentials of living lab environments in urban systems while viewing urban planning as the entire set of transformative practices possible and available in urban contexts. She explores the three main potentials of urban living labs, i.e., their being practice-based innovation environments, their capacity to create cross-boundary arenas where many diverse actors and organizations can interact, and, lastly, their being contexts for new modes of urban activism. This chapter also analyzes some challenges launched by living labs in urban environments and discusses some possible roles for planners who recognize living lab potentials as transformative drivers. Finally, considering the collective (public) experimental perspective introduced by urban living labs, the idea of the city as a laboratory is introduced.

Within the conceptual framework of civic intelligence, **Douglas Schuler** looks at smart cities as the environments for smart citizens to become key actors of their cities (smart citizens). Questioning the equation in the title, "Smart Cities + Smart Citizens = Civic Intelligence," Schuler tries to characterize the type of smart citizens needed to verify the equation. He states that this type of person is more likely to emerge when a stronger democracy is called for, when people and groups feel that they are more valued and have a larger stake in the future. He believes this can happen in laboratory-like environments, but the big challenge that emerges (and must be dealt with) is related to the need to change the mindset of people, combined with the need to widen opportunities for people to collaborate while being engaged (a little at least!) in the definitions of the modes they interact with the city.

The second part, "Complex Participatory Design Processes," describes the idea of complex participatory design as those design processes that operationalize the new vision for experimental policies. Participatory design is here introduced as the most suitable approach for dealing with experimental visions for cities. It is based on activities such as prototyping and testing as experiments that, by their very nature, include the possibility of failing and of continuous changing as key components in city-making.

The section starts with the contribution edited by **Alessandro Deserti**, which offers an overview of the evolution of the relationships between design and the city. The author discusses the main practices of design applied to cities from the early 1970s to the present day. The discussion introduces the different practices of design for cities as a consequence of the radical transformation of the design discipline. Design used to be focused on the small scale, but it has been progressively expanding its fields of interest and application, which completely changed the role that the discipline can play in city transformation. The contribution is complemented with the interesting example of the Milan Design Week, which is used to explain different design practices applied to city design.

The second contribution, by **Francesca Rizzo**, focuses attention on a specific design practice, service design, and how it manifests itself in different city contexts to satisfy new city and citizen needs. The contribution explores the intuition that design can act as an agent of change for public institutions, which are currently facing new and unmet societal challenges that appear to affect cities at different levels. These include the quality of the services offered by municipalities, the way in which public institutions deal with service innovation in conditions of scarce resources, with new phenomena such as social innovation. In the second part, the chapter introduces a design-led project implemented in the framework of the My Neighbourhood European Project, with a double aim, i.e., to experiment service design as a tool for designing innovation in the public sector and to experiment service design as a tool to boost innovation in the culture of a public institution (the municipality of Milan).

The third contribution, written by **Guy Julier**, is about design as a political driver of change in the relations between citizens and cities. It frames the notion of what being a "design city" means. It starts from the idea of design city as framed by neoliberal notions of competitiveness, creativity, and entrepreneurialism. But to go beyond the broad "design city" rhetoric, there are four dominant strategies that cities have used to exploit design objects and economies in order to gain competitive advantage. The strategies are the energetic deployment of urban design to revive the cityscape, the use of place-branding to strengthen and build enthusiasm for its internal and external image, the encouragement of creative industries (through clustering in particular), and the overlaying of cultural planning that boosts reputation in social and cultural capital. Each has been challenged for different reasons and, while there are plenty of examples in which these strategies are still applied, the author suggests viewing them historically in terms of evolution toward other, more recent approaches. The contribution then tries to build a more rounded view of design citizenship, illustrating how currently the two main features of design are developing more complex, sophisticated approaches to design and are embedding them into public sector systems of governance and social innovation.

The third part, "Innovating Governance and Economic Models," analyzes the most evident outcomes of experimental approaches to city-making. These outcomes are explored from the point of view of governance and economic models and give space to a rich reflection on the issue of frugality in both the models.

Revisiting the frugal government concept is the aim of the chapter by **Francesco Molinari**, who first discusses the renewed importance of the concept. Then the chapter provides a working definition for a government to be frugal, i.e., parsimonious in the use of resources, minimizing the costs and maximizing the benefits, and diligent as a good family man. The concept is then further analyzed with its implications for the logic model of government. A discussion follows on the rationale and implications of restructuring public service delivery according to frugal principles and, in particular, the connections between participation, learning, and behavioral change.

The chapter by **Raine Mantysalo** critically reviews the complexities involved in the idea of public-private-people partnership (4P) in urban planning. It does so by questioning how the coexistence of different communication modes can be managed in 4Ps, without losing the partnerships' capability to perform. In discussing this challenge, Mantysalo focuses on the possibility of co-coordination between communication modes through the development of tools and platforms for boundary-crossing communication. For this purpose, the concepts of "boundary object" and "trading zone" are examined, and a tentative analogy is drawn to "urban living labs." In the conclusion, a re-conceptualization of the 4P in urban planning is suggested, as a local trading zone of urban planning.

Jean-Louis Laville introduces the fundamental role of local initiatives and authorities in the perspective of service provision. Laville looks at the local scale as the scale at which groups genuinely invent services thanks to their implicit or explicit perception that there are no suitable answers to their problems. In this way, they create forms of civic entrepreneurship heading toward social enterprises that, importantly, are shaped by solidarity-based services, which are able to hybridize monetary and nonmonetary resources and create forms of work integrating exchange and production rules. Social enterprises do not only create jobs, but they also represent innovative interactions between the economic, social, and political spheres that Laville emphasizes as offering new theoretical inputs going far beyond sectorial constraints.

The need for a new civic economic model is under the focus of **Eugenio Leanza** and **Gianni Carbonaro**. They first discuss the causes and the conditions of the urban unbalances in Europe which are likely to persist in the aftermath of the Great Recession. Leanza and Carbonaro elaborate on the evolving role of cities within the EU and the mounting challenges for cities suffering from shrinkage and economic decline, considering the increasing aging of the European population. They argue that the effectiveness of centrally managed policy instruments to combat these trends is limited. So they make the case for the concept of civic economy enabled by a bottom-up, decentralized policy approach and sustained by innovative urban management practices, which they believe can address at least some of the challenges confronting European cities.

The fourth part, "Experiences," collects four initiatives carried out in very different urban contexts, namely, Sletten in Denmark, Milan and Palermo in Italy, and Aveiro in Portugal.

The Sletten case study, by **Stefan Darlan Boris** and **Peter Gall Krogh**, is both an actual suggestion for novel neighborhood design and a full-scale experimental platform, in which scientists and practitioners with different or no professional background met and collaborated on testing and developing new concepts for the design, establishment, and management of urban forests. In Denmark, the allotment garden movement has diminished over the last decades, but at the same time, there has been growing demand for other types of urban gardening and the ability of people to settle into their surroundings. The Sletten case tells of the people living in Sletten who have been actively involved in the development of the project, but, more than this, it tells the story of how this engagement has led some of

Sletten's inhabitants to establish grazing guilds and animal husbandry, contributing to maintaining the open grasslands despite the lack of any farming experience and facilities.

In the case of Milan, **Valeria Fedeli** discusses the process through which the municipality of Milan has recently started investing in forms of urban agriculture as a way to enhance the active participation of citizens in the production of public goods and to promote unexpected solutions to unsolved urban problems. Urban agriculture in Milan, in this perspective, seems to provide an opportunity for reconciliation between the urban and nonurban realm in a densely built, urban environment and the possibility of introducing new approaches to public policies, urban planning included.

In the case of Aveiro, the authors, **José Carlos Mota** and **Gonçalo Santinha**, present the story of a major urban regeneration project named "The Sustainability Park" (Parque da Sustentabilidade) that was launched in 2009 by the city council. As soon as the project became public, primarily through the media, citizens protested about the fact that they had not participated in the proposal's design, nor had they been informed, to say the very least. This situation prompted a strong civic movement to call the city council's attention to the potential harmful impact of these projects on the local environment and on the daily lives of residents. This contribution describes how movements tried to engage in a series of talks with the politicians involved and bring tacit and codified knowledge to the process.

The case of Palermo, written by **Dino Trapani**, is an example of human smart city (HSC) approach transferred to an extremely weak urban context in southern Italy. The case focuses on the processes boosted through the territorial living lab established in the city to tackle the topic of citizen participation in the solution of the main city challenges. It was based on a model of incremental and adaptive process of interaction, within a co-creative ecosystem, with the aim of improving housing quality. The model was shared by the players involved in the social innovation process. It would not have been possible without the living lab approach and the innovative momentum of design thinking translated into public practice.

Milan, Italy Grazia Concilio
Bologna, Italy Francesca Rizzo

Acknowledgments

The inspiration behind this book on the new association between design and urban planning rests on a long-lasting dialogue with some of our friends. These clever people have been consciously or unconsciously working on the implementation of the productive union between design and cities for years, after decades during which both disciplines developed experiments, cases, and attempts to merge the microlevel of the everyday activities performed by people in their cities with the macrolevel of vision development and durable trajectories for the sustainable growth of cities.

We would like to thank **Jesse Marsh** who, with his continuous practice as a designer active in many territorial projects, has been for us a source of inspiration and a teacher during the projects we developed with him. We also would like to thank **Ezio Manzini** who, with his emphasis on social innovation as a driver for city transformation, has brought to our attention the possibility of starting from the microlevel of collaborative networks as the minimum design unit in urban and territorial environments. Finally we would like to thank all the partners of the Periphéria and My Neighbourhood EU projects that have been the think tank for the development of the human smart city idea pervading this book and guiding our current research.

Grazia Concilio
Francesca Rizzo

Contents

Contributors

Louis Albrechts Department of Architecture, Urbanism and Planning, Catholic University of Leuven, Leuven, Belgium

Stefan Darlan Boris Platform Urbanism and Landscape, Aarhus School of Architecture, Aarhus C, Denmark

Gianni Carbonaro QASER Lab, University College London, London, UK

Grazia Concilio Department of Architecture and Urban Studies, Politecnico di Milano, Milan, Italy

Àlvaro Duarte de Oliveira Alfamicro, Sistemas de Computadores lda, Caiscais, Portugal

Alessandro Deserti Department of Design, Politecnico di Milano, Milan, Italy

Valeria Fedeli Department of Architecture and Urban Studies, Politecnico di Milano, Milan, Italy

Guy Julier Victoria and Albert Museum, London, UK

University of Brighton, Brighton, UK

Peter Gall Krogh Department of Engineering, Head of Design Research, Navitas Inge Lehmanns Gade, Aarhus, Denmark

Charles Landry Comedia, Sheffield, UK

Jean-Louis Laville Conservatoire National des Arts et Métiers, Paris, France

Eugenio Leanza Mandate Services Division, European Investment Bank, Luxembourg City, Luxembourg

Raine Mäntysalo Department of Built Environment, Aalto University, Espoo, Finland

Francesco Molinari Department of Architecture and Urban Studies, Politecnico di Milano, Milan, Italy

José Carlos Mota, Ph.D. Department of Social, Political and Territorial Sciences, University of Aveiro, Aveiro, Portugal

Francesca Rizzo Department of Architecture and Territorial Development, University of Bologna, Bologna, Italy

Gonçalo Santinha, Ph.D. Department of Social, Political and Territorial Sciences, University of Aveiro, Aveiro, Portugal

Douglas Schuler The Evergreen State College, Olympia, WA, USA

Ferdinando Trapani Department of Architecture, University of Palermo, Palermo, Italy

The original version of this book was revised. An erratum to this book can be found at DOI 10.1007/978-3-319-33024-2_16

Part I
Needs for New Durable Visions

Chapter 1
Strategic Planning as Governance of Long-Lasting Transformative Practices

Louis Albrechts

Abstract This chapter argues that new conceptual infrastructures and transformative practices are needed to cope with contemporary environmental and societal challenges. It reflects first on the nature and characteristics of transformative practices that construct images/visions of a preferred innovative structural outcome and opportunities for their implementation. The chapter then deals with the political-economic context and proposes envisioning as a learning process, discussing (among other things) transformative triggers and the power of visions in complex planning contexts. This is followed by a brief description of two cases, Hasselt and Antwerp, which are relevant in terms of transformative actions and visions. Finally, it provides ingredients for more radical strategic planning as the governance of collective affairs, which requires planners to cultivate activist modes of planning and to be more than navigators keeping the ship on course.

Keywords Transformative practices • Strategic planning • Visions

1.1 Introduction

In her book *Collaborative Planning*, Patsy Healey (1997b) understood planning as a governance activity occurring in complex and dynamic institutional environments, shaped by wider economic, social, and environmental forces that structure, but do not determine, specific interactions. By "governance" she meant the processes by which societies and social groups manage their collective affairs (see Healey 2003: 104). Before we can focus on these processes, we have to reflect on the major problems, challenges, and potentials our Western societies, Western cities, and city regions in particular are facing. The challenges are huge and include growing complexity (rise of new technologies, changes in production processes, crisis of representative democracy, diversity, globalization of culture, and the economy), increasing concern about the rapid and apparently random course of development,

L. Albrechts (✉)
Department of Architecture, Urbanism and Planning, Catholic University of Leuven, Leuven, Belgium
e-mail: louis.albrechts@kuleuven.be

G. Concilio, F. Rizzo (eds.), *Human Smart Cities*, Urban and Landscape Perspectives 19,
DOI 10.1007/978-3-319-33024-2_1

the problems of fragmentation, population aging, the dramatic increase in interest (at all scales, from local to global) in environmental issues, the long-standing quest for better coordination (both horizontal and vertical), the returned emphasis on the need for long-term thinking, and the aim to return to a more realistic and effective method. Secchi and Viganò (2011) frame the current condition of cities all over the world as a new urban question that arises in connection with globalizing markets and financial systems. There is a growing awareness that a number of planning concepts (compact cities, livable cities, creative cities, multicultural cities, fair cities, just cities, smart cities) cannot be achieved solely through physical hard planning. There is also awareness of the fact that (in addition to traditional land use regulation, urban maintenance, production, and management of services) governments are being called upon to respond to new demands and to adopt a more entrepreneurial style of planning in order to enhance city and regional competitiveness. This implies abandoning bureaucratic approaches and involving skills and resources that are external to the traditional administrative apparatus. All these developments and challenges serve to expand the agenda. I am fully aware that the problems and challenges are constantly changing and hence resistant to description in terms of fixed categories (see also Chia 1999: 211). To cope with the developments, challenges, and opportunities (see also Sager 2013; Allmendinger and Haughton 2010), planning is in desperate need of a critical debate (which questions the political and economic processes of which existing planning approaches are an integral part (see Sager 2013: xviii)) and a search for new ideas (see Allmendinger and Haughton 2010: 328). This in line with the call by Secchi and Viganò (2011) for new infrastructures: new principles and theories that can act as guides for finding reliable solutions to complex problems related to growing social inequalities, the environmental crisis, and the right to mobility. Which theories, principles, and practices are up to dealing with the (structural) challenges ahead? A growing amount of literature (Healey 1997a, 2006, 2007; Albrechts 2004, 2013, 2014; Albrechts and Balducci 2013; Motte 2006; Balducci et al. 2011) and an increasing number of practices, all over the world, seem to suggest that strategic spatial planning may be considered a possible approach that can deal with the challenges and embed structural change. Despite a certain popularity of strategic spatial planning (see Metzger 2012: 781), one cannot be blind to the critical appraisal of strategic planning. The reviews (see Albrechts 2014) focus on very different registers of the strategic spatial planning approach. Some are related to the ontology and epistemology of strategic planning. Questions are raised on how (and to what extent) the shift from a Euclidian concept of stable entities toward a non-Euclidian concept of many space-time geographies (see Friedmann 1993: 482; Graham and Healey 1999) is reflected in strategic spatial planning. How are the different types of knowledge (tacit/experiential knowledge of local communities versus traditional scientific knowledge) relevant to a relational strategic planning, reflected in strategic plans and actions based on these plans? Economic-political ideological reviews draw a link between the rise of strategic spatial planning and the strengthened neoliberal political climate (see Olesen 2011, 2012; Olesen and Richardson 2012; Cerreta et al. 2010). They fear that the ideal of strategic spatial planning could be easily used to favor the most aggressive neoliberal models of urban and regional development (Cerreta et al. 2010: x; see also Olesen 2011; Sager 2013), and questions are raised as to whether strategic spatial planning practices are able to resist the hegemonic discourses of neoliberalism (see Olesen 2011, 2012). Others attack the militaristic and corporate terminology (see Leal de Oliveira 2000; Adonis Barbieri 2008) of strategic planning. Lastly, there are those who focus on the implementation of the theory in practice, asking whether actually existing practices of strategic spatial planning really follow in line with its normative grounding (see Allmendinger and Haughton 2009, 2010; Newman 2008).

This chapter argues that new conceptual infrastructures and transformative practices are needed to cope with the continuing and unabated pace of change driven by the (structural) developments and challenges. Therefore it starts by considering the nature and characteristics of transformative practices. Secondly, it deals with the political-economic context. Thirdly, it proposes envisioning as a learning process and deals briefly with two cases (the towns of Hasselt and Antwerp). Finally, it provides ingredients for a more radical strategic planning as a contribution to the discussion for new ideas and new conceptual infrastructures. The chapter relies on a selective review of planning literature and the author's experience in practice.

1.2 Transformative Practices: Nature and Characteristics

The transformative agenda is a modern term for structural change that has been discussed by many in the past in the context of planning theory (see Ozbekhan 1969; Schon 1971; Etzioni 1971; Friedmann 1987). Transformative practices focus on the structural problems in society, construct images/visions of a preferred outcome, and how to implement the images/visions. Transformative practices simply refuse to accept that the current way of doing things is necessarily the best way. They break free from concepts, structures, and ideas that only persist because of the process of continuity. It is precisely the discontinuity that forces cities and regions outside the usual boundaries of reasonableness (see de Bono 1992). Transformative practices take decision-makers, planners, institutions, and citizens out of their comfort zones (see also Kotter 1996) and compel them to confront their key beliefs, to challenge conventional wisdom, and to look at the prospects of new ideas and breaking out of the box. Beliefs and expectations matter just as much as reality. Transformative practices must be imagined as being radically and structurally different from the present reality. They focus on new concepts and new ways of thinking that change the way resources are (re)used, (re)distributed, and (re)allocated and the way regulatory powers are exercised. In this way transformative practices become the activity whereby (taking into account structural constraints) that-which-might-become is imposed on that-which-is, and it is imposed for the purpose of changing what-is into what-might-become. Becoming privileges change over persistence and novelty over continuity (Chia 2002: 866). This means a shift from an ontology of being, which privileges outcome and end-state, toward an ontology of becoming, in which actions, movement, relationships, process, and emergence are emphasized (Chia 1995: 601, 1999: 215). So I argue to think in terms of the heterogeneous becoming of institutional transformation, the otherness of institutional outcomes, and the immanent continuity of institutional traces (Albrechts 2010). The transformative invents, or creates, practices – in relation to the context and to the social and cultural values to which a particular city is historically committed – as something new, instead of a solution reached as a result of existing trends.

A number of strong manifestos for structural change have been drawn up (see Albrechts 2005a, b), for reconsidering the absolute faith in economic growth (Misham 1967; Hamilton 2004), for living interculturally (Landry 2000; Sandercock 1998, 2003), for reacting against existing and persistent inequalities (Harvey 2000), and for creating a more sustainable society (Sachs and Esteva 2003). In order to (even partially) implement these manifestos, society needs to mobilize all the necessary resources in such a way that these new ideas develop the power to travel and translate into an array of practice arenas. They have to transform these arenas, rather than merely being absorbed into them. The ideas and ways of thinking that accumulate sufficient power to become routinized may then sediment down into the cultural ground, which sustains ongoing processes and feeds into new processes (Hajer 1995; Albrechts and Liévois 2004; Healey 2005: 147–148, 2006: 532). Until a change truly sticks through its institutionalization into the structure, systems, social norms, shared values, and, most of all, culture, new strategies remain fragile and subject to regression as soon as the stimuli associated with a change effort are removed (Kotter 1996, 2008; Albrechts 1999). Real transformation takes time and dedication and therefore risks losing momentum if there are no short-term goals to meet and actions to celebrate (see Kotter 1996, 2008). Most people will not go on the long march unless they see compelling evidence that the process is producing acceptable results within reasonable periods of time. Indeed, short-term results can build the credibility needed to sustain efforts over the long haul and help to test a vision against actual conditions (see Kotter 1996; Kotter and Rathgeber 2005), but we may not maximize short-term results at the expense of the future. Transformative change rarely occurs in instant revolutions. It is change that actually evolves in many small ways (see Hamdi 2004) to produce an emergent pattern, which, retrospectively, comes together and becomes evident in what history may then describe as a transformative moment (Chia 1999: 212; Healey 2005: 158, 2006: 541).

Transformative processes must be placed within a specific context (economic, social, cultural, and political, as well as in terms of the power structure), place, time, and scale regarding specific issues that are of interest and within a particular combination of actors. This context provides the setting for the process, but it also takes form and undergoes changes in the process. Transformative processes must be rooted in an understanding of the basic processes and constraints that shape places. This must be done fully recognizing the conditions of power, inequality, and diversity. Who benefits from a transformative practice remains a basic question to be asked.

1.3 Political-Economic Context

Recently, a number of authors have argued that strategic planning practices have taken a neoliberal turn. They even link the rise of strategic spatial planning with a strengthened neoliberal political climate (see Olesen 2011, 2012; Olesen and Richardson 2012; Cerreta et al. 2010). They believe that the ideal of strategic spatial planning could be easily used to favor the most aggressive neoliberal models of urban and regional development (Cerreta et al. 2010: x; see also Olesen 2011). As neoliberalism assumes that socio-spatial problems have a market solution (see Peck and Tickel 2002; Swyngedouw et al. 2002; Purcell 2009), its aim was and is to depoliticize the economy (Friedmann 1992: 83) and to subordinate everything to the economic realm and to the sovereignty of the market (Mouffe 2005: 92). In many cities, urban revitalization is presented as an (the?) opportunity to change economic hierarchies and functions within the urban region, creating new jobs and strengthening the city's position in the urban division of labor. Planners and local authorities are lured to adopt a more proactive and entrepreneurial approach aimed at identifying market opportunities and assisting private investors in taking advantage of them (see Harvey 1989; Peck and Tickel 2002; Swyngedouw et al. 2002; Purcell 2009). A democratic deficit emerges as a central element of the neoliberal approach (Purcell 2009:144; Swyngedouw et al. 2002: 573).

Neoliberal policies attempt to create competitive cities and regions by generating investments in major cities and urban regions (Olesen and Richardson 2012: 1692; Swyngedouw et al. 2002). Such investments (projects) have become a key component of a neoliberal shift from distributive policies, welfare considerations, and direct service provision toward more market-oriented and market-dependent approaches aimed at pursuing economic promotion and competitive restructuring (see Swyngedouw et al. 2002: 572). Questions have to be raised whether strategic spatial planning practices are able to resist the hegemonic discourses of neoliberalism (see Olesen 2011, 2012). Everyday practice proves that it is enormously difficult to enact, by sheer force, the structural changes that are needed to tackle the problems and challenges our society is facing today. So the challenge seems to be to look for ways to help citizens, politicians, planners, and civil society as a whole to discuss and reflect upon the challenges, opportunities, and the new conceptual infrastructures and transformative practices needed. We have to be aware that we cannot confront complex dynamic realities with a language designed for simple static problems (Senge 1990). Hence there is the need for ways of thinking and for tools and instruments that help society cope with change in a dynamic environment (see also Winch 1998).

1.4 Envisioning as a Learning Process for Transformative Practices

Envisioning is the process by which individuals (or preferably groups) develop visions of future states regarding them, their organization and their city, that are sufficiently clear and powerful to arouse and sustain the actions necessary for (parts of) these visions to become a reality (see Goodstein et al. 1993). The term "visions" refers to images of possible futures with some implicit or explicit commentary on why civil society should try to build these futures. Visions or frames of reference are not just out there, waiting to be discovered – on the contrary, we have to construct them. This is not a linear but rather a dialectic (backcasting and forecasting) process. Visions themselves may not be seen as static descriptions of futures. They must comprehend and portray the dynamic nature of development, structural constraints, and changing challenges and contexts. Without an appropriate vision, a transformation effort can easily degenerate into a list of confusing, inconsistent, and time-consuming projects that move in very diverse and often incompatible directions or else go nowhere at all (Kotter 1996). In a change process, a vision serves several purposes. It points, in a very specific way, to the critical issues and challenges ahead, creating a sense of urgency among as many actors as possible (Kotter 2008). It astonishes and confronts our most deeply ingrained beliefs about what is important and why (see Hames 2007). It fights complacency (Kotter 1996, 2008; Kotter and Rathgeber 2005) and reveals how things can be different and truly better by shifting the unthinkable into the realm of the possible (see Hames 2007). It motivates actors to take action in a specific direction and it helps to frame the actions of different actors. It allows interdependent actors to work with some degree of autonomy and yet not trip over each other. The actors involved will only make sacrifices when they truly believe that a transformation is possible or feasible and that the potential benefits of change are attractive (Kotter 1996). It must become clear to the citizens, politicians, and planners how specific policies, behavior, and attitudes help. The vision describes what a city may look like in the future and it must appeal to the long-term interests of the actors – and the less privileged as well – who have a stake in the city. The ambition expressed in the vision needs to be high enough that it cannot be accomplished through business as usual (Kotter 2008). It must comprise realistic but at the same time daring strategies or actions. Therefore the vision must be focused, but also flexible, in order to allow alternative responses in the light of changing conditions. A vision must be sufficiently clear and easy to understand and communicate. This does not mean that its basis is not solid or that it is not backed by relevant information and analytically sound thinking. Effective visions seem to be rooted in sensible values that resonate deeply with civil society. The process very much involves getting in touch with who we are and what we care about (see also Kotter 1996). Envisioning does not claim to eliminate uncertainty through the making of predictions – it seeks instead to work as well as possible within the context of uncertainty and to enable the actors to open up the spectrum of possibilities. Envisioning is, above all, a state of mind (imagination and anticipation) that leads to behavior (hope and will) (see Godet 2001: 8). In the final analysis, we must come back to what "is" if we want to present ideas and concepts that are solid, workable, and of testable value. To have to these ideas, we need both the solidity of the analysis and the creativity of the design of alternative futures. Since envisioning is also the journey and not just the destination, and as visions are so central to transformation and so all-invasive, envisioning cannot be confined to a single actor or institution in the process. The values and images of what a society wants to achieve are not generated in isolation, but are socially constructed and are given meaning and validated by the traditions of belief and practice. They are reviewed, reconstructed, and invented through collective experience (see Ozbekhan 1969 but also Elchardus et al. 2000: 24; Foucault 1980: 11; Hillier 1999). Indeed, the individual self is only a by-product of perpetually shifting constellations of relations, never a fundamental stable unity in its own right (Chia 2003: 969). This implies that the objects that society engages with are not things, but are more an interconnecting set of shifting relations in which action is undertaken based on an awareness of divergent understandings (e.g., different professional codes or local norms) and not only on the identification and use of material assets (Chia and Holt 2006: 649). Therefore I consider envisioning to be a collective process that concerns futures for which a wide variety of actors are themselves responsible. This means their visions are more than wish lists – they involve commitment to the realization of visions through practice (Friedmann 1987). Visions provide actors with views of the future that can be shared, a clear sense of direction, a mobilization of energy, and a sense of being engaged in something important (see Goodstein et al. 1993). One main challenge in this respect is how to shift a what's-in-it-for-me culture to one of civic-mindedness, which sees the current challenges as opportunities. The use of relevant cases could be helpful in this respect. The first case presented here draws on the development of creative transport planning, on creative local governance, and on the role of a key person – a champion, in the terminology of Bryson (1995), which allows for structural change. The second case draws on the use of images as instruments to reveal the characteristics and quality of the city of Antwerp and its possible futures.

1.5 Broadening the Scope of Places

1.5.1 Creative Governance in Tackling a Transport Problem (See Albrechts 2005b)

Hasselt, a regional city in the Northeast of Belgium, is a major commercial and service center with a population of 69,000. Like many cities, it has suffered from the mounting costs of externalities caused by automobile travel, e.g., accidents, traffic jams, and environmental problems. These externalities have had a negative impact on the livability and appeal of the city. Combined with other factors, they have resulted in a decreasing number of inhabitants. In the 1990s, the new local government (a coalition of socialists, greens, and conservatives with a charismatic new mayor) was placed in a dilemma: to build a third ring road or to completely reverse the existing transport policy. The socialist party organized meetings with local residents. In these meetings local traffic proved to be an important issue. In the meantime, a temporary free shuttle bus service was introduced to compensate citizens for the nuisance caused by major local roadworks. This shuttle proved an enormous success. Although the intense discussions of the 1970s (see Bologna) about free public transport were on the decline and even seemed to be disappearing, the mayor launched the pioneering idea of introducing free public transport for citizens and visitors alike in the entire urban area. This places the problem within a broader relational perspective than simply the internal travel patterns of the city's own citizens. This was just one action within the context of a much larger strategy of 22 actions to be carried out in close cooperation with residents, companies, schools, public bodies, etc. The reasoning behind the idea of free public transport was that a considerable shift from car to public transport would make the construction of the third ring road unnecessary and that the first ring road could even be rebuilt. When, in discussions with the public transport company, the mayor found out that only 9 % of the overall cost of public transport was covered by the sale of tickets, he immediately offered to compensate the company for this loss. The savings resulting from not constructing the third ring road would more than offset the subsidies for transit services, thus leading to a positive financial net effect. The radical reconstruction of the first ring road narrowed the car lanes and improved the facilities for cyclists and pedestrians (a 9 m wide pedestrian area bordered by a double row of trees), which considerably improved the livability and overall quality of the urban environment. The results between mid-1997 (when the scheme was introduced) and 2002 have been mixed. On the one hand, there has been an astonishing 1200 % increase in the numbers of public transport passengers and an increase in the number of bus routes from one to nine. On the other hand, the numbers of cyclists have decreased. However, the most important results have been the strengthening of the social tissue and the fact that the elderly became more mobile, that the discourse on public transport turned very positive, and that the extreme right-wing party has not gained a foothold in the city council (in sharp contrast with other similar cities).

In this case, a problem (major road congestion) has been turned into an asset, i.e., free public transport. The problem of congestion was viewed from a different perspective. Indeed, instead of the traditional engineering logic of "more traffic = more roads," the logic of the pedestrians, of the elderly, of public transport, and of the overall livability of the city was introduced. The mayor thought of a solution – free public transport – that no one else had thought of. The costs of constructing a third ring road were turned into net benefits, despite the subsidies paid to the transport company. The mayor enhanced the livability of the city by linking it to the traffic problem. This strengthened social and political capital as citizens and local politicians took more pride in their city, which became a best practice case, attracting visitors from all over the world, including governments, students, all kinds of specialists in transport, etc. This case also illustrates the impact of a leading person. Although the context was not very innovative (a traditional socialist party and an equally traditional engineer-led public works department), the mayor managed to make people think about new ideas and new solutions. The project resulted in a landslide election victory for the mayor and his party. A change in the coalition after the 2012 local elections in a harsh neoliberal context provoked a shift in the city's policy on mobility.

1.5.2 Imagining the City: Antwerp

Antwerp, with 450,000 inhabitants, is the largest city in Belgium. Other than the official, more strategic planning-oriented approach, the city draws its inspiration (equally and even more intensively) from an approach that emerged within the architectural/urbanism discipline on land use regulation and a new generation of strategic (mainly urban) projects, which applies to the revival of run-down parts of cities and regions, such as the French "Projet de Ville," in particular (see Secchi 1986; Motte 1994; Masboungi and De Gravelaine 2002). In addition to traditional sectoral and technical knowledge, design operations provide a specific and original body of knowledge (see Viganò 2010). In this approach urban design, as an instrument to analyze and to read places, uses images and representations to reveal the qualities of existing spaces and places and their possible futures. The active creation of images became a key component of the planning culture in Antwerp.

In their book *Antwerp: Territory of a New Modernity*, Secchi and Viganò (2009) – the authors of the Antwerp structure plan – introduce the idea of the image as a forceful synthesizing tool for understanding and designing urban space. The image in their interpretation relates to the human capacity of understanding and imagination and to human presence and consciousness, suggestive knowledge conveying, and projecting quality. In this way they are simple keys to a collective imagination and memory. Secchi and Viganò construct images (water city, eco city, rail city, harbor city, porous city, villages and metropolis, and megacity) for Antwerp as part of different performative practices in which interpretations and strategic choices about the city are being made.

Secchi and Viganò (2011) zoom in on the porosity concept in Antwerp to illustrate the complexity of the city. This concept holds an interesting innovative perspective for interpreting the interaction between morphology and societal dynamics. It refers to the built and non-built environment, to different social groups, and to the different morphologies of the city. Porosity regards different ways in which different city users (not only human but natural as well) use urban space and move within the city. Porosity is not a static condition: it is related to different phenomena that can modify the way urban space reacts, over time, to practices and movement, pressure, and abandonment. It is part of a renewed concept of mobility, which weaves together the social and physical dimensions. The image of a porous city reflects a permeable and accessible city, where many central places are endowed with their own recognizable identities, disseminated throughout the city and connected by a capillary infrastructure network, the lower network, by a network of shopping streets, and by surface public transport. The image of the porous city can be used as a stimulus for local renovation, to enhance the public domain. It also deals with the relationship and the sharing of public and private space (see Secchi and Viganò 2009). The generic policies in Antwerp regarding the porous city focus on the creation of infrastructure and facilities for conviviality and sociability, guaranteeing the quality of life and work in the city by maintaining a sufficient amount of open space in high-density areas, by reusing vacant areas and empty buildings and by creating a micro-fabric of links for bicycles and pedestrians, and by maintaining a social and functional mixture that can be differentiated in the various neighborhoods. All this contributes to a strategy for the construction of urban space. As a spatial plan, the official structure plan in Antwerp focuses mainly on the spatial (say physical) aspects of porosity. This may lead to an incomplete understanding of space – hence the challenge to expand and enrich the concept of porosity with a sociocultural dimension and consider this dimension as an integral part of the concept. Related concepts such as permeability, connectivity, and accessibility have a potential in this respect. Permeability could go beyond the link with the geometry of the city and refer to permeability for different cultures, social groups, where accessibility could not only focus on the real possibility of moving from one point to another but also on the real possibility of moving socially and economically. Connectivity could not only be linked to physical barriers but also to social, economic, and cultural barriers. Permeability, connectivity, and accessibility acknowledge the existence of barriers and aim to overcome them. Moreover, opening the concept of porosity toward the sociocultural provides an opportunity to reconnect planning in Antwerp to the long-standing social tradition of the city.

The two cases illustrate that breaking out of the box and the use of images are instrumental to broaden the scope of possible futures and strengthen the quality of a city. This is in line with the introduction, where I argued that planning needs to look for new ideas that are up to dealing with the challenges ahead and add to new conceptual infrastructures. In the next paragraph, I reflect on what this means for strategic planning and for planners.

1.6 What Does This Mean for (Strategic) Planning and Planners?

It implies a type of strategic planning that provides direction without destination and movement without prediction, tackles problems, raises awareness, unravels and resists the influence of international neoliberal ideologies on planning theory and planning practices in cities, meets challenges, broadens the scope of the possible (see Zizek 1999: 199 about the art of the impossible; Hillier 2007), encourages hopes and dreams, appeals to values (equity, social justice), provides a frame for decisions, and challenges existing knowledge, conventional wisdom, and practices (see also Hillier 2002, 2007; Brand and Gaffikin 2007; Healey 2010; Metzger 2012). It needs an arena, i.e., a space of deliberative opportunities in Forester's (2010) terms, an open dialogue in which a plurality of interests and demands, opinions, images, conflicts, different values, and power relationships are addressed. In these arenas actors may reflect on who they are and what they want and in this way articulate their identities, their traditions, and their values.

1.6.1 Strategic Planning as Governance of Collective Affairs

If we reflect upon the city in which we live our life, we will be able to discover layers of stakes (Healey 1997b: 69, 91–92, 2006: 542) that consist of existing but perhaps unconscious interests in the fate of our city. Hence comes the plea for strategies that treat urban territory not just as a container in which things happen, but as a complex mixture of nodes and networks and places and flows, in which multiple relations, activities, and values coexist, interact, combine, conflict, oppress, and generate creative synergy (Healey 2007: 1). A challenge in contemporary governance – and by extension in planning – consists in the dialectic between movements that seek democratization, collective decision-making, and empowerment of citizens on the one hand and the established institutions and structures that seek to reabsorb such demands into a distributive framework on the other (see Young 1990: 90). It entails a political struggle between different visions of justice: justice as distribution – presuming a consumer-oriented, possessively individualist conception of people – and justice as enablement and empowerment, presuming a more active conception of people (Young 1990: 15–38). A crucial element in this respect is the way in which people are excluded or included in planning processes and the way the relationship between people, technologies of government, and norms of self-rule (Roy 2009) is organized. If concerns of some groups in society (especially the weak groups) cannot be tackled within the preconceived level of government, new practices will have to be invented. Problematically, a broad range of these relationships is being compressed into a one-fits-all concept of citizen participation, which does not seem to provide the deeply sought equal and reciprocal relationship between the state and (all) citizens. So, for strategic planning to be successful, a key task is to explore who has a stake in an issue. The question concerning who is to be considered a stakeholder in a particular context or situation is not only an epistemological challenge but also a fundamentally ontological issue (Metzger 2012: 782). Therefore strategic planning looks for an arena, a platform that organizes the relationship between (all) actors in a more open and equitable way and where actors can articulate their identities, their traditions, and their values. In different contexts and in different intellectual traditions, this search led to a coproduction approach and engagement between the state and (all) citizens that were more likely to be successful (for references see Albrechts 2013), with citizens as a part of action, not its object (Friedmann 2005), and as a combination of a needs-based and rights-based approach (Albrechts 2013). In a world where actors are interdependent and have a (implicit) reason to engage with each other, coproduction is considered as an engine of change that may make a difference between systems working and failing. This assumes that transformative potential lies in the very multiplicity of tensions and stresses of the relational complexity of coproduction processes, creating all kinds of fissures and cracks which can be opened up to create and enlarge moments of opportunity for new ideas (Healey 1997b, 2006: 540) and new conceptual infrastructures.

Coproduction is constructed as an inclusive and multivocal arena, grounded in a deeper understanding of the complex dynamics of urban and regional relations (see Healey 2006: 541) where value systems can be articulated, where local and scientific knowledge can be combined on an equal base, where shared strategic conviction can grow, and where conflicts are reframed in a less antagonistic manner. It recognizes that knowledge is always partial and sometimes partisan and that the search for enhanced knowledge is endless rather than exhaustive (Brand and Gaffikin 2007: 293). It is inclusionary (for those in and outside the system) and intentional to secure political influence and to change the status quo with specific projects and policies. It provides an interaction between the delivery of public goods (strategies, policies, projects) and building strong, resilient, mutually supportive communities, i.e., coproduction as a political strategy. In a coproductive form of governance, deliberation takes place through a lot of face-to-face interaction in real time (see Friedmann 1993: 482). Face-to-face dialogue (see Legacy 2010: 2706) allows to acknowledge the role of the emotional and the personal dimension, expressed in the narrative that allows the whole person to be present in negotiations and deliberations (Sandercock 2000: 6). The dialogic process is itself transformative in the relations among the participants, creating a "sensing together" rather than the conventional consensus, whereby antagonism can be domesticated into agonism (Hillier 2002: 289). Only under such conditions can policy be designed, not for citizens but by citizens in their role as policy users (see Brand and Gaffikin 2007: 290). Using coproduction in a strategic planning process offers alternatives, stimulates critical reflection, is noncoercive, and is capable of reflecting particular experiences with more universal principles (equity, social justice) and issues (sustainable development, spatial quality) (see Brand and Gaffikin 2007 294). Coproduction as a central concept embodies a social science perspective (see also Cahn 2000: 29) and strengthens the socio-spatial character of the strategic planning process. It is looked upon as a process of becoming, with outcomes which must be well informed, just, and fair. In this way coproduction is part of a much broader shift that is emerging across all the sectors and most obviously in those fractures between and in public and private.

As an alternative for institutionalized and taken-for-granted practices and routines, the added value of incorporating coproduction (conceived as a political strategy) for the selection of problems, discussing of evidence, of strategies, of justice or fairness, and of the nature and scope of desired outcomes, is that it is conceived as a learning process, which permits a plurality of problem definitions, ambitions, and ways to achieve it for those inside and outside the system. Coproduction introduces into the neighborhood, city, or region new identities and practices that disturb established histories. In this way coproduction may strengthen the local organization base of citizens – more specifically the urban poor – and increase their capacity to negotiate successfully with the state (see Mitlin 2008: 340) and other powerful actors. In its own politicization, it may have the capacity to hold a mirror to the process of neoliberalization revealing its real character, scope, and consequences (see Peck and Tickel 2002: 400). It is focused on developing socio-spatial imaginations that reinvent modernism's activist commitments to the construction of places and to the construction of an inclusionary governance system. In this way it includes not only the views of the most articulate or powerful but also the views of those who have been systematically excluded by structural inequalities of class, gender, and religion (Sandercock 1998: 65), and as a learning process, it gains an emancipatory potential. This implies that strategic planning may not be locked within the interstices of the state and the powerful actors in society (Friedmann 2011). Coproduction requires a change to the status quo and is conceived as a creative task of generating collective becomings underwritten by a democratizing ethos (see Metzger 2012: 794), and there are no technical rules and norms according to which coproduction processes are to be conducted (see Roy 2003, 2009; Mitlin 2008; Watson 2011). With strategic planning, as conceived here, people are being asked to construct their own governance institutions (see Healey 1997b: 209). Good governance is possible only if institutions are allowed some margin for self-governance of a form appropriate to their particular tasks, within a framework of financial and other reporting (ONeill 2002: 58). As classical institutions are still endowed with substantial powers, it is clear that redistributive policies also need to be framed in more general redistribution and regulatory policies at higher-scale levels (see also Swyngedouw et al. 2002).

1.6.2 Impact for Planners

Because strategic planning is not only instrumental, it may not be reduced to a set of neutral procedures. The implicit responsibility of strategic planners can no longer simply be to be efficient or to function smoothly as a neutral means for obtaining given and presumably well-defined ends. The normative dimension inscribed in strategic spatial planning is of an ethical nature, as it always refers to values (equity, social justice) and specific practices (see also Healey 2010 for the crucial normative foundation of strategic spatial planning). Without the normative, we risk adopting a pernicious relativism where anything goes (see Ogilvy 2002; Metzger 2012: 793).

The second case draws on the creation of images as contextual, conscious, and purposive actions to represent values and meanings for the future of a city. This provoked a shift from analysis, which seeks to discover a place that might exist, toward design (in its broadest sense), which creates a place that would not otherwise exist. This is similar to Habermas' (1996) idea of knowing (understanding the challenges and the options available) and steering (the capacity to take action to deal with the challenges).

Strategic planners have an active but not a dominant role in a planning process (see Metzger 2012). As active generators of conditions of collective becoming (Metzger 2012: 793), they must be more than navigators keeping the ship on course and they are necessarily involved with formulating that course. In this line of reasoning, they are not just looking for existing articulated interests – they are actively involved in broadening the scope of the possible and in articulating and bringing to the table the interests that can be of a collectivity that may yet become (Metzger 2012: 794). This implies an activist mode of planning (see Sager 2011, 2013 for an overview of activist modes of planning). For planners working in the system (government planners), an equity type of planning (Krumholz and Forester 1990) open to local knowledge and where citizens and the disadvantaged become an equal part of the action seems suited. For planners working outside the system (NGOs, community organizations), only a radical type of planning (for references see Sandercock 1998: 97–104) makes it possible to work for structural transformation of systemic inequalities. In this way strategic planning is undoubtedly a political process.

1.7 Epilogue

The two cases discussed above make it clear that creativity, breaking out of the box, the use of different languages, and logics are instrumental in broadening the scope of possibilities. Strategic spatial planning in this chapter is not looked upon as the ultimate model, neither as a panacea for all challenges and all problems, nor as a new ideology preaching a new world order. It is not meant as a substitute but as a complement for other planning tools (statutory planning, urban design). It is presented as a method for creating and steering a (range of) better future(s) for a place (see also Ogilvy 2002). Its focus on becoming produces quite a different picture compared to traditional planning in terms of plans (strategies versus master plans or land use plans), type of planning (providing a framework and a justification for specific actions versus technical/legal regulation), type of governance (government-led versus government-led but negotiated form of governance), and content (vision and actual actions that accept the full complexity of a place while focusing on local assets, socio-spatial quality, a fair distribution of the joys, and burdens).

As illustrated in the Hasselt case, planning is not just a contingent response to wider forces but is also an active force in enabling change. It also shows that planning cannot be theorized as though its approaches and practices were neutral with respect to class, gender, age, race, and ethnicity (Albrechts 2002; Sandercock 1998). The Antwerp case makes it clear that combining an urbanist discourse with a more social planning discourse deepens the scope of some concepts. The recent evolution in Hasselt illustrates that the capacity of a spatial planning system to deliver the desired outcomes is dependent not only on the legal-political system itself but also on the conditions underlying it (see Olesen and Richardson 2012: 1690; see also Needham 2000 for success factors). These conditions, including political, societal, cultural, and professional attitudes toward spatial planning (in terms of planning content and process) and the political will on the part of the institutions involved in setting the process in motion (and in keeping it going, which is much more difficult), affect the ability of planning systems to implement the chosen strategies.

A strategic spatial planning process based on coproduction acknowledges that some forms of strategic spatial planning may tend, in the long term, to reinforce the current status quo because they seek to resolve conflict, eliminate exclusion, and neutralize power relations rather than embracing them as the very terrain of social mobilization (see Purcell 2009: 155). Therefore it is necessary to mobilize sufficient discursive counterpower to challenge prevailing powers and to go beyond the recycling of established discourses, concepts, and practices.

Because the values, interests, views, ideas, and policies from actors are different, strategic spatial planning involves choices and hence it inevitably works in a context of conflicts, clashes between the different actors. As most strategic planning processes are non-statutory processes, questions are raised about the legitimacy of such processes.

References

Adonis Barbieri R (2008) From global to local: Erechim 100 plan and city marketing in the periphery. Paper presented at the 44th Isocarp congress

Albrechts L (1999) Planners as catalysts and initiators of change. The new structure plan for Flanders. Eur Plan Stud 7(5):587–603

Albrechts L (2002) The planning community reflects on enhancing public involvement. Views from academics and reflective practitioners. Plan Theory Pract 3:331–347

Albrechts L (2004) Strategic (spatial) planning reexamined. Environ Plan B Plan Des 31:743–758

Albrechts L (2005a) Creativity in and for planning. DISP 169(3):14–25

Albrechts L (2005b) Creativity as a drive for change. Plan Theory 4(3):247–269

Albrechts L (2010) More of the same is not enough! How could strategic spatial planning be instrumental in dealing with the challenges ahead? Environ Plan B Plan Des 37(6):1115–1127

Albrechts L (2013) Reframing strategic spatial planning by using a coproduction perspective. Plan Theory 12(1):46–6

Albrechts L (2014) Ingredients for a more radical strategic spatial planning. Environ Plan B (Forthcoming)

Albrechts L, Balducci A (2013) Practicing strategic planning: in search of critical features to explain the strategic character of plans. DISP 49(3):16–27

Albrechts L, Liévois G (2004) The Flemish Diamond: urban network in the making. Eur Plan Stud 12(3):351–370

Allmendinger P, Haughton G (2009) Commentary. Environ Plan A 41:2544–2549

Allmendinger P, Haughton G (2010) The future of spatial planning-why less may be more. Town Country Plan (July/August):326–328

Balducci A, Fedeli V, Pasqui G (eds) (2011) Strategic planning for contemporary urban regions. Ashgate, Farnham
Brand R, Gaffikin F (2007) Collaborative planning in an uncollaborative world. Plan Theory 6(3):282–313
Bryson JM (1995) Strategic planning for public and non-profit organizations. Jossey-Bass, San Francisco
Cahn E (2000) No more throw-away people: the co-production imperative. Essential Books, Washington, DC
Cerreta M, Concilio G, Monno V (eds) (2010) Making strategies in spatial planning: knowledge and values. Springer, Dordrecht
Chia R (1995) From modern to postmodern organizational analysis. Organ Stud 16(4):579–604
Chia R (1999) A rhizomic model of organizational change and transformation: perspective from a metaphysics of change. Br J Manag 10:209–227
Chia R (2002) Time, duration and simultaneity: rethinking process and change in organizational analysis. Organ Stud 22(6):863–868
Chia R (2003) From knowledge-creation to the perfecting of action: Tao, Basho and pure experience as the ultimate ground of knowing. Hum Relat 56:953–981
Chia R, Holt R (2006) Strategy as practical coping: a Heideggerian perspective. Organ Stud 27:635–655
De Bono E (1992) Serious creativity. Using the power of lateral thinking to create new ideas. Harper Business, New York
Elchardus M, Hooghe M, Smits W (2000) De vormen van middenveld participatie. In: Huyse L (ed) Hetmaatschappelijk middenveld inVlaanderen EdsMElchardus. MHooghe VUB Press, Brussels, pp 15–46
Etzioni A (1971) The active society: a theory of societal and political processes. Collier Mac Millan, London
Forester J (2010) Foreword. In: Cerreta M, Concilio G, Monno V (eds) Making strategies in spatial planning. Springer, Dordrecht, pp v–vii
Foucault M (1980) The history of sexuality. Vintage, New York
Friedmann J (1987) Planning in the public domain: from knowledge to action. Princeton University Press, Princeton
Friedmann J (1992) Empowerment. The politics of alternative development. Blackwell, Oxford
Friedmann J (1993) Toward a non-Euclidian mode of planning. J Am Plan Assoc 59(4): 482–485
Friedmann J (2005) Globalization and the emerging culture of planning. Prog Plan 64:183–234
Friedmann J (2011) Insurgencies: essays in planning theory. Routledge, London\New York
Godet M (2001) Creating futures. Economica, London
Goodstein L, Nolan T, Pfeiffer J (1993) Applied strategic planning. McGraw-Hill, New York
Graham S, Healey P (1999) Relational concepts of space and place: issues for planning theory and practice. Eur Plan Stud 7(5):623–646
Habermas J (1996) Normative content of modernity. In: Outhwaite W (ed) The Habermas reader. Polity, Cambridge
Hajer M (1995) The politics of environmental discourse. Oxford University Press, Oxford
Hamdi N (2004) Small change. About the art of practice and the limits of planning in cities. Earthscan, London
Hames R (2007) The five literacies of global leadership. Jossey-Bass, San Francisco
Hamilton C (2004) Growth fetish. Pluto Press, London
Harvey D (1989) From managerialism to entrepreneurialism: the transformation in urban governance in late capitalism. Geogr Ann Ser B 71(1):3–17
Harvey D (2000) Spaces of hope. University of California Press, Berkeley
Healey P (1997a) The revival of strategic spatial planning in Europe. In: Healey P, Khakee A, Motte A, Needham B (eds) Making strategic spatial plans. U.C.L. Press, London, pp 3–19
Healey P (1997b) Collaborative planning: shaping places in fragmented societies. Palgrave Macmillan, Basingstoke

Healey P (2003) Collaborative planning in perspective. Plan Theory 2(2):101–123
Healey P (2005) Network complexity and the imaginative power of strategic spatial planning. In: Albrechts L, Mandelbaum S (eds) The network society: a new context for planning? Routledge, New York, pp 146–160
Healey P (2006) Relational complexity and the imaginative power of strategic spatial planning. Eur Plan Stud 14(4):525–546
Healey P (2007) Urban complexity and spatial strategies. Towards a relational planning for our times. Routledge, London
Healey P (2010) Making better places. The planning project for the twenty-first century. Palgrave Macmillan, Basingstoke
Hillier J (1999) What values? Whose values? Ethics Place Environ 2:179–199
Hillier J (2002) Shadows of power. Routledge, London
Hillier J (2007) Stretching beyond the horizon: a multiplanar theory of spatial planning. Ashgate, Aldershort
Kotter P (1996) Leading change. Harvard Business School Press, Boston
Kotter P (2008) A sense of urgency. Harvard Business School Press, Boston
Kotter P, Rathgeber H (2005) Our iceberg is melting. St Martin's Press, New York
Krumholz N, Forester J (1990) Making equity planning work. Temple University Press, Philadelphia
Landry C (2000) The creative city: a toolkit for urban innovators. Earthscan, London
Leal de Oliveira F (2000) Strategic planning and urban competition. Plan Netw 143(September/October):11–13
Legacy C (2010) Investigating the knowledge interface between stakeholder engagement and plan-making. Environ Plan A 42:2705–2720
Masboungi A, De Gravelaine F (2002) Projets urbains en France. Editions du Moniteur, Paris
Metzger J (2012) Placing the stakes: the enactment of territorial stakeholders in planning processes. Environ Plan A 45:781–796
Misham E (1967) The costs of economic growth. Staple Press, London
Mitlin D (2008) With and beyond the state – coproduction as a route to political influence, power and transformation for grassroots organizations. Environ Urban 20(2):339–360
Motte A (1994) Innovation in development plan-making in France 1967–1993. In: Healey P (ed) Working paper 42. Department of Town and Country Planning, University Newcastle upon Tyne, pp 90–103
Motte A (2006) La notion de planification stratégique spatialisée en Europe (1995–2005). Puca, Lyon
Mouffe C (2005) On the political. Routledge, London/New York
Needham B (2000) Making strategic spatial plans: a situational methodology! In: Salet W, Faludi A (eds) Revival of strategic planning. Royal Netherlands Academy of Arts and Sciences, Amsterdam, pp 79–90
Newman P (2008) Strategic spatial planning: collective action and moments of opportunity. Eur Plan Stud 16(10):1371–1383
Ogilvy J (2002) Creating better futures. Oxford University Press, Oxford
Olesen K (2011) Strategic spatial planning in transition: case study of Denmark. PhD thesis, Department of Development and Planning, Aalborg University
Olesen K (2012) The neoliberalisation of strategic planning. Paper presented at the Aesop 26th Annual Congress, Ankara, 11–15 July
Olesen K, Richardson T (2012) Strategic spatial planning in transition: contested rationalities and spatial logics in 21st century Danish spatial planning. Eur Plan Stud 20(10):1689–1706
ONeill O (2002) A question of trust. Cambridge University Press, Cambridge
Ozbekhan H (1969) Towards a general theory of planning. In: Jantsch E (ed) Perspectives of planning. OECD, Paris, pp 45–155
Peck J, Tickel A (2002) Neoliberalizing space. Antipode 34(3):380–404
Purcell M (2009) Resisting neoliberalization: communicative planning or counter-hegemonic movements? Plan Theory 8(2):140–165

Roy A (2003) City requiem, Calcutta; gender and the politics of poverty. University of Minnesota Press, Minneapolis
Roy A (2009) Civic governmentality: the politics of inclusion in Beirut and Mumbai. Antipode 41(1):159–179
Sachs W, Esteva G (2003) Des ruines du développement. Le Serpent à Plumes, Paris
Sager T (2011) Activist modes of planning: a systematic overview. Paper presented at WPSC, Perth
Sager T (2013) Reviving critical planning theory. Routledge, New York/London
Sandercock L (1998) Towards cosmopolis. Planning for multicultural cities. Wiley, Chichester
Sandercock L (2000) When strangers become neighbours: managing cities of difference. Plan Theory Pract I 1(1):13–30
Sandercock L (2003) Cosmopolis ll. Mongrel cities in the 21st century. Continuum, London
Schon D (1971) Beyond the stable state. Maurice Temple Smith, London
Secchi B (1986) Una nuova forma di piano. Urbanistica 82:6–13
Secchi B, Vigano P (eds) (2009) Antwerp, territory of a new modernity. Sun, Academia, Amsterdam, p 248
Secchi B, Vigano P (2011) La Ville poreuse, Un projet pour le Grand Paris et la métropole de l'après-Kyoto. Edition MetisPresses, Genève
Senge P (1990) The fifth dimension. Doubleday, New York
Swyngedouw E, Moulaert F, Rodriguez A (2002) Neoliberal urbanization in Europe: large-scale urban development projects and the new urban policy. Antipode 34(3):542–577
Viganò P (2010) A territorial project. In: Meijsmans N (ed) Designing for a region. Sun, Amsterdam
Watson V (2011) Planning and conflict – moving on. Paper presented at WPSC, Perth
Winch G (1998) Dynamic visioning for dynamic environments. J Oper Res Soc 50:354–361
Young I (1990) Justice and the politics of difference. Princeton University Press, Princeton
Zizek S (1999) The ticklish subject. Verso, London

Chapter 2
Urban Living Labs: Opportunities in and for Planning

Grazia Concilio

Abstract This chapter explores some of the most significant potentials of Living Lab environments in urban systems while viewing urban planning as the entire set of transformative practices possible and available in urban contexts. It explores three main potentials of Urban Living Labs, i.e., their being practice-based innovation environments, their capacity to create cross-boundary arenas where many diverse actors and organizations can interact, and, lastly, their being contexts for new modes of urban activism. This chapter also analyzes some challenges launched by Living Labs in urban environments and discusses some possible roles for planners who recognize Living Lab potentials as transformative drivers. Finally, considering the collective (public) experimental perspective introduced by Urban Living Labs, the idea of the city as a laboratory is discussed.

Keywords Urban Living Labs • Practice-based innovation • Urban experiments

2.1 What Is a Living Lab?

The term "Living Lab" refers to a conceptual research and development approach developed in computer science at the end of the 1990s. The concept of Living Lab originates from William Mitchell (MIT Media Lab) and was promptly adopted in human-computer interaction focused on user-centered design principles. It easily evolved to a more radical and human-driven approach to design and to participatory design in particular. The concept has been adopted very rapidly in many sectors moving from product development to service design, so it became the key concept in many smart city initiatives and projects as well.

There are many definitions for what a Living Lab is. Svensson et al. (2010) and Molinari (2011) consider a Living Lab to be a methodology, as well as an organization, an environment, and/or a system. Eriksson et al. (2005) describe a Living Lab as "a user-centric research methodology for sensing, prototyping, validating and refining complex solutions in multiple and evolving real life contexts." Another

G. Concilio (✉)
Department of Architecture and Urban Studies, Politecnico di Milano, Milan, Italy
e-mail: grazia.concilio@polimi.it

G. Concilio, F. Rizzo (eds.), *Human Smart Cities*, Urban and Landscape Perspectives 19,
DOI 10.1007/978-3-319-33024-2_2

relevant definition is the one provided by ENoLL (the European Network of Living Labs; see http://www.openlivinglabs.eu/): a Living Lab is a methodology for user-driven innovation and the organizations that primarily use it. Boronowsky et al. (2006) see the Living Lab as a setting of shared resources and opportunities focused on finding answers to problems and helping one another to achieve their goals. In these environments people (experts and non experts) interact and work in productive ways, designing and implementing cooperative and joint experimental activities that result in collective learning and shared understanding.

Generally, all the definitions view users/consumers of services and products as contributors and cocreators engaged in an innovation process. In Living Labs collaboration and learning happen within the complexity of real life and the produced results have the nature of common goods.

As environments where the general public is involved in an innovation process, Living Labs are intended to operationalize the Quadruple Helix innovation model (Arnkil et al. 2010) instead of the Triple Helix model proposed by Etzkowitz in 1998. The latter considers three actors collaborating with each other in order to create or discover new knowledge, technology, products, and services. The actors are universities (providing science-based knowledge and technologies), governments (formulating policies to support innovation), and firms (developing and marketing products). The Quadruple Helix innovation model introduces a fourth actor enriching, and in coherence with, the idea of open innovation developed by Chesbrough (2003) and further enriched by von Hippel (2005). The additional actor is the general public, which has been traditionally excluded from innovation processes, so this model requires the creation of new innovation protocols and environments.

Most of the existing Living Labs deal with new technologies in everyday contexts as they are used by people (often referred to as "users") to achieve their goals. They are interaction systems where innovation occurs in real-life contexts through collaborative interaction, design, and experimentation in special environments that are open in nature (Dutilleul et al. 2010; Leminen et al. 2012). In these contexts, people with different experiences, skills, expertise, and motivations explore innovative tools and solutions by using and interacting with them. In this way they improve them as well as discovering new ideas, expanding their knowledge, and exploring different, new ways of acting (Følstad 2008).

Because they look promising, Living Labs have been considered a key approach for innovation in urban systems (see the Joint Programming Initiative Urban Europe 2013–2014). Much still needs to be understood about the dynamics introduced by these environments into urban systems. The main aim of this chapter is to explore some of the most significant potentials of Living Lab environments in urban systems while viewing urban planning as the entire set of transformative practices possible and available in urban contexts (Albrechts 2006; Van der Broek 2008). The discussion starts by relating the Living Lab concept to the specific character and needs of urban innovation. Next, three main potentials of Urban Living Labs are explored, i.e., their nature of practice-based innovation environments, their capacity to create cross-boundary arenas where many diverse actors and organizations can interact, and, lastly, their nature of contexts for new modes of urban activism. In the second part, some challenges launched by Living Labs in urban environments and some possible roles for planners are discussed. Finally, some reflections are developed considering the collective (public) experimental perspective introduced by Urban Living Labs toward the idea of the city as a laboratory.

2.2 Living Labs and Urban Innovation

Living Labs are described as open innovation systems. When referring to innovation processes, the "openness" concept generally referred to is the definition given by Chesbrough (2003) for the corporate environment: "Open innovation is a paradigm that assumes that firms can and should use external ideas as well as internal ideas, and internal and external paths to market." When referring to urban or territorial innovation, the openness concept appears embedded in the nature of socio-spatial environments, but it is harder to achieve because it challenges the governance models and practices of such environments through relevant and complex principles like active democracy, distributed decision-making, and power delegation. Living Labs are labeled as territorial context-driven environments, in which user-centered research and development activities are carried out in an open innovation ecosystem rooted in a territorial/urban context (for Territorial Living Lab, see the seminal work by Marsh (2008) and also Concilio and Celino (2012)). Similarly, Urban Living Labs are defined as creative communities of people producing innovation at urban level with the support of a number of methods and tools, which help cocreate value out of the experience of interaction between the citizen/customer and the public/private service provider (Concilio and Molinari 2014). Both Territorial and Urban Living Labs are able to activate concurrent research and innovation processes within a public-private-people partnership and to keep those processes alive (Von Hippel 1986; Chesbrough 2003; Pallot 2009).

In a recent literature review on urban and territorial Living Labs, Wallin (2015) identifies three types of Living Labs in socio-spatial environments:

- Living Lab as a technology-driven research environment
- Living Lab as a testing environment for know-how and tools
- Living Lab as an arena for self-organizing groups

The first kind is the one most closely related to the original idea and definition. It is a research environment developed in a real-life context to codesign, prototype, and test new technologies. In this type of Living Lab, the role of the user is that of an observed subject and only slightly that of a citizen engaged in the cocreation of ideas and breakthrough scenarios. In the second type of Living Lab, methodologies and technologies are tools for reaching the users and for transforming the real urban environment with the users. The goal of such a Living Lab is to encourage the users to develop and produce urban/territorial artifacts and to cocreate urban/territorial spaces. The third type of Urban Living Lab is also an approach to the development of urban/territorial systems but it is not created under the planning/design intention of an external actor. It is instead born and created by the self-organized collective intention of a community in its everyday life condition (Wallin and Horelli 2010, 2012). Very often it is related to the collective need to resolve the problems the community experiences most.

One of the underlying ideas in the Living Lab approach is that people's ideas, experiences, and daily needs should be the starting points for innovation (Bergvall-Kåreborn et al. 2009). This principle represents the main characterizing element of Living Labs in urban environments, because the importance of daily life experience makes citizens and their interaction with urban life a fundamental contribution to the innovation processes, regardless of the followed approach. As Simon Marvin recently stated (quoted in Gibson 2014), urban complexity inevitably requires experimental research approaches. Many authors (Wallin 2015; Concilio and Molinari 2014; Hou 2010) showed that cities are experiencing experimental laboratory initiatives that are challenging policy makers, planners, and researchers and asking for a change in the way urban systems are approached. According to Marvin, this shift partly represents a response to a fragmented discourse on sustainability, which still appears far from being an urban condition and yet asks for urgent responses. The concept of urban laboratories is emerging more and more as an approach to speed up socio-technical innovation involving an unprecedented multiplicity of actors in coproduction processes. City centers, neighborhoods, and peripheries are increasingly being transformed into experimental platforms to explore the needs of users as residents and citizens (Juujärvi and Pesso 2013). These laboratories host experiments and practice-based innovation processes with diffuse and heterogeneous knowledge production that aims to address urban problems of varying and increasing complexity. The recent shift from traditional to open innovation makes the importance of experiments and practice-based innovation more crucial when dealing with urban systems, because it drives toward a relevant modification in the role assigned to citizens. From being "users" of whatever service cities can supply, citizens become coproducers of innovation together with all the other different actors. Innovation in such processes is strictly related and consequent to collective learning, a fundamental mechanism behind experimental and practice-based innovation:

> An actual innovation is an innovation whose importance it has been possible to demonstrate, or which has for other reasons achieved a certain degree of acceptance and legitimacy and, as a consequence, a certain degree of [adoption]. Thus, what constitutes an actual innovation – rather than a creative idea, invention or discovery – can only be determined on the basis of its practice. (Ellström 2010: 29)

Living Labs, as experiments and practice-based innovation environments, represent opportunities for cities to develop and use new urban routines, products, and services. This is obvious in some sense if we consider that, before a novelty (a new service, plan, procedure, etc.) is accepted, it has to demonstrate its contribution to the accomplishment of some goals, which can happen only in the practice of the novelty itself. This process is difficult enough in the closed organizational environments of business companies or public institutions and it is far more challenging and complex when the novelty has to enter the daily life of urban systems.

2.3 Practice-Based Innovation

Cities are considered service platforms (Walravens 2013). The way people interact with urban service platforms is shaped by service practices and it is at the level of the practices that innovation can occur. Everyday practice is the main resource for Living Labs.

As innovation is the process of turning knowledge and ideas into value, urban innovation can be seen as a learning process acting on (and transforming) practices in urban collective environments. According to Ellström (2010), this learning process is related to practice-based innovation, i.e., the development and use of new methods, routines, products, or services.

This kind of innovation occurs at a local level. Entering the urban environment, it appears to be quite common to acknowledge something as an innovation if it is perceived as such at the local scale at which it is developed or implemented, even though its novelty may appear limited in a broader context (Miles 1964; Pettigrew and Fenton 2000). Such "localism" of innovations often entails reinterpretations, or new interpretations, involving more or less innovative elements in the practice (Sevón 1996). This may be in line with the view that innovations are based on new combinations of elements that are already well known, and perhaps also applied, yet have not been previously linked together – at least not in the local context (Pettigrew and Fenton 2000). Still, they represent innovations that are locally able to create benefits and to accomplish specific goals through the creation and use of practice-based knowledge.

Practice-based knowledge is produced continuously in situated actions, as people draw on their cultural background but mainly on their direct experience in a social setting. Practice-based knowledge is embedded and distributed among the several, multiple, diverse actors who interact with it. It does not exist independently of the social action; it is situated in the collective action and it is not the same for all the actors involved. Practice-based knowledge is collective, since no single person can know all the heuristics or principles involved or possess all the necessary experience (Cook and Brown 1999). This is generally true, yet it is especially true when dealing with urban services, which are infrastructure involving highly complex social processes. Knowledge embedded in practice is crucial to innovation but cannot be transferred like a physical object that is packed up and moved. This has a relevant consequence when looking for innovation, because innovation has to occur in strict dependency with practices.

Urban Living Labs are the environments where this knowledge can become a resource for innovation; they are experimental environments for reflecting on practices. Here experiments in and through the practices activate a common ground for knowledge creation toward the (re)definition of the practices themselves. This is possible because, in Living Labs, knowledge that is embedded in the daily practices (Itami refers to them as "operations," 1987) of urban services becomes operational as an innovation resource. Urban Living Labs, as complex social processes of urban service practices that interact with complex social processes of service (re)design (Dougherty 1992; Leonard-Barton 1995), are environments in which important practice-based innovations are produced.

It is in Urban Living Labs that people collectively enact the three kinds of activities for generating practice-based knowledge for innovation:

> interweaving designing and using (or, routinely doing their particular designing activity in terms of its impact on using, not apart from it), participating in the whole flow of designing and using (or, routinely doing their part in terms of its relationship to the whole, not apart from the whole) and reflecting in action (or routinely iterating from emergent knowing to articulating that knowledge). (Dougherty 2004: 43)

These activities constitute a common ground of social action, so people can engage in situated learning and make sense of what they learn in other occasions of daily life in urban environments.

Many definitions of innovation are available, such as "central to many definitions of the innovation concept is the idea that an innovation relates to some form of specific change that is new (at least locally) and that leads to what is in some sense a better accomplishment of goals at the system level (the local unit or the larger organization/system of which it is a part)" (Ellström 2010: p. 28). The practice-based innovation perspective explains the first requirements of the definition used by Ellström, novelty. It appears relevant with regard to the second additional requirement as well: for something to be accepted as an innovation, it must be possible to demonstrate that it contributes to the accomplishment of the goals of an organization (or system). This last requirement is usually possible at a later stage of an innovation process. Local innovation, such as the one taking place in Urban Living Labs, entails a proposed change that is in some sense new. However, it has not yet been demonstrated that this change will be able to contribute to the accomplishment of the goals of the system (or be of importance to a group or operation in some other way), and it cannot therefore be legitimized in terms of its results or importance to others.

An innovation at urban scale is an innovation whose importance (as a contribution to goal accomplishment or in some other way) has been possible to demonstrate or which has achieved a certain degree of acceptance and legitimacy for other reasons.

Looking at cities as complex systems where diverse practices characterize the interaction with the urban service platform, we can imagine that innovation (service innovation) comes from a networked complex system of local, small-scale interdependent experiments, where (1) new or modified practices are developed and combined together to solve problems and to respond to needs, (2) new practices or practice modifications are developed (designed, prototyped, and tested) in action at the scale at which problems are practically experienced, (3) there is increased awareness of the interscalar interdependency of urban actions and practices, and (4) learning at an urban-scale system becomes operationally possible thanks to experiments being considered public/collective goods.

2.4 Cross-Boundary Environments

In Living Labs the interplay among a mixture of different world views, value systems, and organizations is handled by continuously crossing the existing boundaries between them (Aldrich and Herker 1977). In this sense Living Labs can be considered as cross-boundary innovation contexts. They are open innovation contexts where cocreation between a multiplicity of actors and stakeholders happens on the boundaries between their sectors, knowledge, and organizations. They remind us of what Nonaka and his colleagues (2000) define as "ba," i.e., knowledge-creative contexts and shared spaces for knowledge creation. Here, in these cross-boundary contexts, the interfaces between sectors and organizations become reciprocally permeable, making specific, sectorial, organizational knowledge and resources generate new collective learning and action throughout the innovation process.

Their cross-boundary nature makes them particularly suitable for dealing with urban problems and complexity. In the complexity of urban systems and problems, the (often very high) difficulty of interaction among various organizational realities and urban actors, each with different roles, routines, knowledge domains, expertise, and fields of action, is crucial. Ad hoc environments, being cross-boundary in practice, are necessary for making such wealth become a collective value, while creating innovation from and upon distances and diversities (Fischer 2005) rather than keeping them as separate and independent entities.

Working across the boundary, several artifacts such as scenarios, prototypes, design drawings, and other types of objects can be used to move the interaction toward a collective dynamic. The literature defines these artifacts as *boundary objects*.

The theory of boundary objects, originally developed by Susan Leigh Star (1990), has been applied and used to analyze and support the interactions that take place and the objects that people create and use in the context of crossing the boundaries of different organizations and communities, generically described as "social worlds" by the literature. Boundary objects can cross the boundaries between multiple and diverse social worlds as well as being used within and adapted to/by many of them at the same time (Star and Griesemer 1989, p. 408). They adapt to specific needs but are also robust enough to keep their identity across boundaries, so links between different social worlds become possible.

While the concept was initially originated referring to the studies of various scientists, the concept has acquired a wider and more general value: as suppliers of "interpretive flexibility", boundary objects operate as supports for different kinds of mechanisms (such as heterogeneous translations) or as knowledge integration mechanisms, mediating entities in coordination processes of experts and non experts in many diverse interaction contexts (Worrall 2010). Therefore, the concept acquired relevance in many different domains, from sociology to medical science, from information technology to architecture, and also in urban planning and design disciplines (Aibar and Bijker 1997; Milanovich 2006; Mantysalo et al. 2011; for a broad review of the concept use see Trompette and Vinck 2009). The recognition of their relevance is not limited to them as tools, or objects; it is also a result of the richness and complexity of the social dynamics and mechanisms they activate or support when in use.

Symbols, metaphors, concepts, organizations, systems, cultures, and ICTs residing on the boundary of different communities or systems have often been recognized and/or used as boundary objects in many different Urban Living Lab experiences. Although no exhaustive analysis of existing Urban Living Lab experiences has been fully developed, it is possible to identify some common elements among them and consider the nature of the boundary objects produced. Among these, some can be recognized to be more suitable for innovation in urban environments.

Collective/Public Spaces. Collective/public, and in general physical, spaces can support the activation and life of a cross-boundary innovation when they act as boundary objects (Concilio and Moro 2015). Usually they work better as such when they are open to being interpreted as both private and public environments, in order to increase the number of actors acting in these spaces and/or feeling responsible for them.

Technologies. Technologies used in collective practices and activities play a key role in mediating and enabling cross-boundary interactions (Forgues et al. 2009). This happens much more when rules and protocols of ICT-based interaction practices are embedded in the ways these objects are adopted and used. This happens more and more with technologies open to different uses and functions, as they allow the interpretation, change, adaptation, and reconfiguration of collaborative patterns.

Strategies. Strategies are traditional planning tools and have often proved to act as boundary objects to interact across boundaries (Spee and Jarzabkowski 2009). The best strategy tools are those considered transparent and easy to use (Clark 1997; Stenfors et al. 2004), because they do not require specialist knowledge or skills and are largely supportive and usable in interactions among different actors. The more simple and transparent they are, the more can they be considered appropriable (Campbell 1997) and effective in overcoming the boundaries.

Visions, Scenarios, and Perspectives of the Future. The future is well known to be something people can easily find the way to agree on as it just "is," by nature. Often trans-boundary processes are driven by perspective making and taking (Boland and Tenkasi 1995), both representing the first steps of learning, where knowledge creation is built and rebuilt for shared understanding and communication. Celino and Concilio (2010) have discussed the role of low-structured scenarios as tools for activating and aligning collective actions.

Symbols and Metaphors. Symbols and metaphors are traditional tools in planning practices. They play a significant cross-boundary role: they allow the creation of a shared language and the coordination among different communities as well. A brand, a leader, an anecdote, and a climate evaluation form all of these and many other potential symbols can be effective (Michalski 2006). Metaphors and symbols are often used in planning. Metaphors are often used in strategic planning to provide a synthetic representation of the scenario adopted by a strategic plan. They are usually kept general and lightly structured on purpose, in order to gain in ability to cluster diverse and dispersed actors, interests, and resources.

Boundary objects are not acting across the boundaries alone – *subjects* are relevant because they have different roles. These subjects represent the power engines of Living Labs; they are the activators of (and also caregivers for) process dynamics. They are the engagers of new actors and energizers for experimental activities and theirs is the main responsibility for the life of the cross-boundary context.

Johansson and his colleagues (2011) focus on the *brokers*: they enforce, activate, and take care of relations that can be observed and/or developed between the objects and the activities or situations composing the process itself. Brokers act as third parties with respect to the organizations involved, although they may belong to one of them. They can be intentionally or casually in charge of process caring and keep the cross-boundary context active by being intermediaries of the objects between the involved actors. Crucial for brokers are the ability and aptitude to recognize the opportunities for linking objects to specific situations.

Also relevant in this perspective are actors acting as *boundary crossers*, key mobilizers of social capital, who provide leadership and cross the boundaries between the different involved organizations and worlds through their relationships. They understand the organizations' different capacities and can lead actions to build and use these capacities inside cross-boundary environments (Kilpatrick et al. 2014).

2.5 Collective Spheres of Urban Activism

Urban Living Labs represent urban environments where new collectives can emerge. Often they are environments that are activated by citizen opposition movements and needs for responses or they are enabled by specific government or management intentions acting at the urban scale. Following the framework developed by de la Pena (2013), Urban Living Labs can be described as forms of urban activism where *autogestión*[1] can prevail and yet be supported or flanked by *reform* approaches. De la Pena describes these environments as in tension between power distribution

[1]The original term *autogestión* used by de la Pena is preferred here to the English term *self-management* in coherence with the author's choice; see de la Pena's explanation at page 2, 2013.

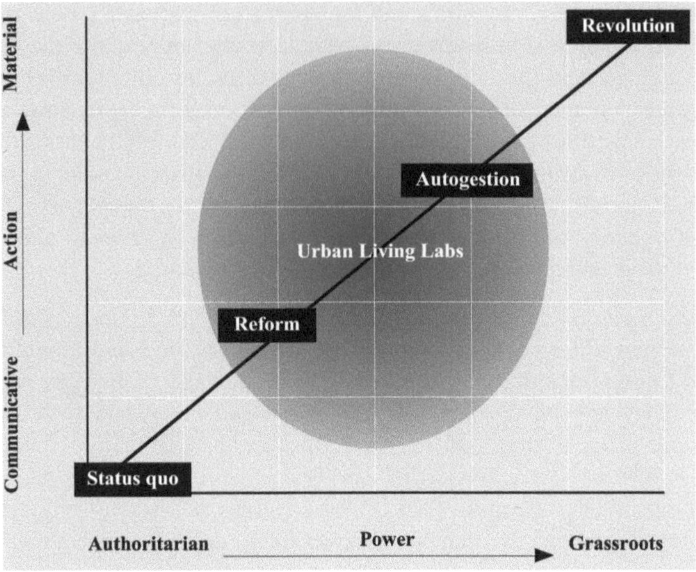

Fig. 2.1 Urban Living Labs and modalities of urban activism (Modified from de la Pena 2013: 19)

(variable, ranging between authoritarian and grassroots) and action aim (variable, ranging between communication and material place making). Urban Living Labs work exactly inside this tension (Fig. 2.1). Here citizens communicate with different intensities, and their actions in the urban space can be considered a measure of the power they are gaining or using for city-making and transformation.

Still, in coherence with de la Pena, these environments are characterized by both official and unofficial modes of collective action in urban spaces. The official (formal) modes include most of the forms (procedures, methods, and practices) of citizen participation being used and experimented all over cities and towns by urban managers, policy makers, and planning practitioners. They are activated within predefined governance structures and rarely include space making and transformation roles for citizens.

Unofficial (informal) modes are autonomously created by different groups or organizations being more and more active in city-making. Groups and organizations (often informally) create grassroots initiatives, which are located in the cities or urban neighborhoods, i.e., at the very microscale. These initiatives are driven by a collective intention and represent a new form of activism, which is less explicit in protesting (as in the traditional meaning of this term) and more operational and proactive in fixing or changing urban space. These too can be viewed as Urban Living Labs: citizens look for autonomous answers to growing needs, especially when public administration fails to accomplish its mission. People enter the public sphere experimentally, not just to demonstrate their dissent, but by taking charge of substantial portions of it, they create experimental opportunities for new forms of governance.

In both cases, formal and informal modes create the framework of collective spheres of urban activism in which a sort of action convergence is working. In these collective spheres, it is crucial to be involved in the "action" (be it communication, material city-making, or something else). Acting, and being available to act inside a collective, is the mode for citizens to reveal a collective agreement that is not to be intended as a consensus (in terms of values, beliefs) – it is something like a temporary availability to act, contingent to the opportunity offered (in the case of formal modes) or to the created or self-organized situation (in the case of informal modes). It is limited to the specific action planned at a precise moment in the urban context. This convergence is temporary and ephemeral when referring, among other things, to the whole process it belongs to: a person verifies his/her availability at each step, and each step represents a new collective condition. This is typical of experimental environments where consensus itself, or any of its forms, can be part of the experimental process. "I do something with others" is valuable here and now and its contingency is a relevant feature of the collective sphere. It represents the symptom of a form of liberty, which is possible independently (more easily) of the verification of collective and shared values, beliefs, and principles. A person has something at stake in joining and acting with other people for as long as he/she considers acceptable.

Despite their temporary and ephemeral features, these environments can slowly develop durable phenomena of civic engagement. In some cases they remain experiments – collective experiments – whose outcomes are considered common goods although they do not root into the urban environment. In other cases, the outcomes slowly (sometimes even rapidly) scale up into urban phenomena or movements, attracting a high number of actors and transforming ephemeral collective spheres into civic movements (Concilio and Molinari 2015).

2.6 Promises and Dilemmas of Urban Living Labs

The conditions for a multiplicity of actors and competences to design and test a wider range of possible solutions exist in Urban Living Labs. Although being embedded in urban reality, Urban Living Labs are in fact environments where a certain degree of independence, typical of experimental settings, is guaranteed by the small scale of the intervention (among other things), so an accurate focus on problem solution and a larger use of creative and innovative human resources are possible.

As open innovation environments, Living Labs (and Urban Living Labs in particular) are a sort of temporary system, which groups a mixture of different knowledge actors and specialist competences for the achievement of a certain goal related to the problems experienced in the urban environment. The traditional barriers to changes and innovations are more effectively circumvented due to the small scale of the processes activated in Urban Living Lab environments. Living Labs are not usually activated in urban environments as large-scale and permanent changes. Because of their experimental nature, they are localized, concentrated attempts to find suitable solutions to specific problems. Here the produced changes are rarely irreversible, as they create a small disturbance in the urban system as a whole.

Moreover, experimented solutions use small amounts of resources and are frugal (see Molinari in this book) from two different points of view. To begin with, they are developed with resources available in the specific problem contexts and do not require relevant additional economic or physical resources (citizens are more prone to mixing resources than professional designers). Secondly, they are developed and tested in spaces of proximity, localities. They are situated and consequently frugal in dimension and do not require large investments. This frugality adds to these solutions being reversible and effective in urban environments.

Consequently, Urban Living Labs represent the environments where a large variety of actors can negotiate while interacting with urban managers and city officers, which are the contextual conditions for space and service innovation. In these environments the specialization and competition of and among urban governmental units are more easily overcome thanks to the strict focus on specific, often localized, problems. Conditions can be developed in Living Labs for different organizations to work across boundaries and this can also simplify the way municipal units interact and cope. The activation of Urban Living Labs has no clear and specific origin and their outcomes do not either; they are environments where organizational or regulation constraints challenge actions softly and lightly and so enable a broader and more effective interaction and collaboration.

In spite of the several advantages that can be envisaged when considering Urban Living Labs as drivers for urban innovation, they also pose a recurring set of *dilemmas* for urban practices and governance, which have implications when analyzing possible associated planning perspectives. One of such recurring dilemmas is related to the sphere of autonomy required by Living Lab practice in urban environments. It creates tension with Urban Living Labs being embedded in the urban setting, where routine conditions are necessary for the coordination of the urban system as a whole. While independence and autonomy are necessary for creativity and innovation, coordination in complex collective systems (such as urban systems) is essential to ensure that the knowledge gained inside a specific (suburban) environment (specifically, inside an Urban Living Lab) is stored as a learning resource for the urban system as a whole (Sydow et al. 2004).

Related to this, there is another recurring dilemma: innovation processes in Urban Living Labs are often carried out while challenging normative and regulative systems working on their boundaries or even, in some cases, with no respect for such boundaries. In several occasions (see the wide literature on the institutional and normative impact of Living Labs), the need to relax regulation and normative systems has been underlined in order to guarantee the existence of conditions for creativity and innovation as well as liberty and sustainability (Moroni 2015). However, regulations and norms are the basic infrastructure for urban management, because they ensure equal access to urban resources and services and their changes and modifications need to be verified against the public and collective interest.

Urban Living Labs may represent an occasion for positive and effective changes but, at the same time, they require an important monitoring and management ability of their instantiations.

Finally, Urban Living Labs also introduce a dilemma related to what Warner (2011) defines the "Swiss cheese" effect. Referring to service delivery carried out by or through clubs, Warner reflects on the fact that many clubs are supported by governments. "While this works well for those inside the club, for the remaining public sector the result is a local government sector that looks a bit like a Swiss cheese, where the clubs are the holes and the remaining public sector is the connective tissue that holds the system together. But, in a world of clubbed and fragmented local government, that connective tissue may become thinner than ever. Local government risks becoming a Swiss cheese dried out around the edges of all the holes (clubs), not a vibrant connective tissue" (p. 157). Surely Urban Living Labs do not represent "holes" in local government, but each innovation they introduce may be seen as a microfracture and a local discontinuity in public services. If they grow in numbers and diversity, they may undermine and weaken the city as a whole by fragmenting its identity and production process.

2.7 Opportunities for Planners

Urban Living Labs show important potentials for urban transformation and innovation. They are very productive environments where cities can experiment new modes of city-making and where many different actors can explore new roles in city-making.

Some of the most important characteristics of these innovation environments, which appear relevant for cities especially when looking for responses to daily life problems, have been discussed. These obviously present some critical issues but they also suggest to planners opportunities to be captured and transformed into values for the planning process. In this chapter three different characteristics of Urban Living Labs have been analyzed: the practice-based nature of the innovation they produce, their functioning across boundaries between different organizations and institutions, and their being very special collective spheres of recent and productive urban activism. These characteristics explain the capacity of Urban Living Labs to overcome some of the most important inertia of contemporary urban lives and can be considered productive opportunities to enrich and empower planning activities. Considering the three characteristics, some possible roles for a planner aware of the potentials of Urban Living Labs in urban transformation have been identified and discussed.

Enabler of Experimental Practices. Planners who share the idea that cities are integrated platforms of/for urban practices are deep and acute observers of urban life and are able to recognize the physical stages and collective encounters that have the strategic power to create and experiment new practices. Such planners view experiments of urban practices as new forms of appropriation of the city (Brasil 2010) and within such experiments they learn together with the city while creating new forms of citizenship.

Laboratory Manager. Planners, being the drivers of urban transformation, greatly value the role of urban experiments as public laboratories for collective learning and production of knowledge as a public good. In urban laboratories, ideas, solutions, and new practices are disseminated and cross-fertilized while interfering with and contributing to the construction of our cities and urban societies. Within this perspective a planner looks at a city as a permanent laboratory, continuously testing and innovating solutions. His/her role becomes the one of a laboratory manager whose main goal is to discover new modes of city-making for, and together with, the citizens.

Boundary Spanner. Planners can facilitate the cross-boundary work by behaving as boundary spanners, i.e., as gatekeepers or brokers. These roles are relevant for creating and keeping the conditions for the collaboration in the practice of multiple actors and organizations. This requires a certain level of legitimacy to influence the development of practices or activities which are not urban routines. An effective boundary spanner mobilizes attention, addresses conflicting interests, and enables learning by combining different practices from different organizations. Following Ancona and Caldwell (1992) and their four roles related to boundary spanning, a planner with boundary spanning abilities may act as a scout (scanning the environment for information and resources), as an ambassador (engaging in lobbying, impression management, and protecting the Living Lab actors from noncoherent pressures), as a coordinator (coordinating and negotiating with the context outside the LL), and as a guard (preventing inadequate distortions from the goals).

Boundary Object Handler. In the planning practice, two different situations may be faced: (i) if a situation exists that can be valuable for "the collective," the creation or adoption of an effective boundary object can achieve a wider engagement in the situation, transforming it into a cross-boundary context, and (ii) when a new action or situation is needed as a new chance for the collective, boundary objects can be created or adopted to activate it and make it suitable for different organizations, sectors, or domains. In both cases it is clear that two alternative conditions for boundary objects may exist: boundary objects can be designed for the purpose of entering or activating activities or for situations considered useful for the collective. Alternatively, boundary objects can already exist (see the discussion on designated boundary objects and boundary objects-in-use by Spee and Jearzabkowski 2009) as self-produced by the involved organizations, in which case they only need to be recognized by the planner in order to acquire larger value in the cross-boundary interaction.

Interim Manager. Exploring the possible existence of human boundary objects, Zdunczyk (2006) underlines the importance of individuals operating on the margins of a collective, plural action. These marginal actors have the virtue of partial and simultaneous membership in different social worlds and this allows others to view them in different categories. For a planner who is in charge of a precise urban transformation task, acting as an interim manager is an effective way to perform the role of a marginal actor. An interim manager may be seen in different ways by different parties: as a traditional manager by those who are his/her direct dependents, as a consultant by the involved municipal managers, and as a colleague by the members of the engaged professional networks. The existence of an interim manager would improve the practical (and political) capacities of an Urban Living Lab.

Coupler and Decoupler. Planners can be seen as couplers and/or decouplers of Living Lab processes and results from or with their contexts, e.g., by referring or not referring to the Living Lab practice as typical or innovative in the complex urban organization and structure. This activity is strictly related to the ability to perceive, recognize, and capture the situational condition of the Living Lab process. It may need to develop a deep resource interdependence and therefore ask for a coupling, attaching approach to the urban context (see Sahlin-Andersen and Soderholm 2002, for a reflection on these themes). In different situations it may need to be more independent from the urban context in order to gain in innovation and learning opportunities. With respect to these changing needs, the planners' role would be the one of ensuring these conditions in coherence with a larger and shared urban vision or scenario.

Urban Activist. Planners are increasingly required to identify, capture, and care for opportunities that ensure citizens have more local control over urban transformations and spaces. This implies the capacity to negotiate the dynamic terrain among institutions, professionals, and public bodies but also to be available to act as urban activists, i.e., people who work toward the changes they believe in. Planners (being urban activists) are doers, driving forces, or impulses of urban changes. They approach changes as results of reforms or by promoting the reforms needed to maintain changes that have an insurgent nature. Planners as urban activists are agents of transformation, privileging material over communicative action (de la Pena 2013), thus endorsing, promoting, stimulating, as well as securing, defending, and preserving it in urban environments.

Warden of a City Vision. Urban innovation requires guidance from a strategic perspective, i.e., from a perspective able to keep the coherence with a strategic vision. This allows an additional possible role to be played by planners – they can be seen as guardians of the city vision that enables its coherent growth toward a differentiated but coordinated identity. In a sense a planner may support the functioning of the connective tissue described by Warner (2011) as the necessary gathering of and holding together all the disparate innovations activated in the cities. Keeping a vision able to value and coordinate innovation initiatives would avoid excessive fragmentation and the consequent dispersion of the innovation potentials. Coordinated within a city vision, innovation initiatives could become resources for learning at a scale higher than the one of the individual innovation.

2.8 Closing Remarks: Toward "Lab Cities"

Living Labs represent a more or less recent fashionable concept that is being used and referred to everywhere user- or community-driven innovation is considered crucial. This innovation model has been already largely explored and investigated in the ICT, service design, and production management domains. Many initiatives of the European Commission have been inspired by, or oriented to, the implementation of this approach (all smart city initiatives or initiatives related to mobility or energy efficiency in urban environments are relevant to this book). Being too "fashionable" may be the reason why urban planning scientists are only recently showing an interest in this phenomenon and have mainly been distracted from considering and fully exploiting its potential benefits for urban systems.

Far from considering Living Labs as "the" solution to the critical, complex problems cities are facing everyday, this chapter has discussed some of the most evident potentials offered by the Living Lab approach for altering consolidated city-making practices and protocols. The Living Lab's most important characteristic is represented by the main idea at the core of its definition, i.e., the idea of exploring and developing solutions by experimenting in real-life environments. This idea is extremely powerful because it suggests that experimental attempts, ideas, observations, results, failures, and successes all belong to a collective learning process that transforms them into common goods. Solutions (practices, services, behavioral changes, etc.) developed inside Living Lab environments already belong to the cities as they have been developed by the cities themselves.

Cities able to capture the potentials of Living Labs may be envisioned as experimental platforms where innovation is possible. Such cities are:

- Able to activate practice-based knowledge production in collective environments, not being kept as isolated environments, being organized instead as nodes of a learning network able to make their experiments the experiments of others and to conceive knowledge not in terms of a competitive advantage tool but rather as a public good.
- Able to learn internally but also externally, by experimenting with other cities that share the same problems; city networks are not effective learning environments yet because they are traditionally based on experience sharing (best practice model), whereas urban experimentation networks in which cities can learn with other cities while experimenting are possible (nothing of the same is enough and many experiments are needed to define integrated and sustainable solutions).
- Aware of new civic engagement models being experimented all over the world, also fed by the innovation insurgences emerging in the urban environments in response to citizen needs; innovation insurgences are spontaneous experimentation labs that many cities are offering for new forms of urban making (Concilio and Molinari 2015).
- Aware of the growing demand for new citizenship models (Harvey 2003) and are willing to develop them with citizens, with those who do not yet share the right to a livable city; citizenship models based on cocreated urban environments and dynamics are possible.

- Capable of directing investments toward opportunity creation (i.e., experiments) rather than pre-developed solutions, with investments being larger in number but smaller and reversible (see Molinari's concept of frugality in this book) and limited to specific experimental scales; investments are not implementing existing solutions; they are creating innovation capital instead.

Urban Living Labs, with their approach to innovation, enable a vision of cities as experimental environments in which new modes for city-making can be explored and developed. They enable a vision of the cities as collective laboratories where the cities (as a whole) themselves can really be transformed into collective artifacts.

References

Aibar E, Bijker W (1997) Constructing a city: the Cerda Plan for the extension of Barcelona. Sci Technol Hum Values 22(1):3–30

Albrechts L (2006) Bridging the gap: 'From Spatial Planning to Strategic Projects'. Eur Plan Stud 14(10):1487–1500

Aldrich H, Herker D (1977) Boundary spanning roles and organizational structure. Acad Manag Rev 2(2):217–230

Ancona DG, Caldwell DF (1992) Bridging the boundary: external activity and performance in organizational teams. Adm Sci Q 37(4):634–665

Arnkil R, Järvensivu A, Koski P, Piirainen T (2010) Exploring quadruple helix outlining user-oriented innovation models. Työraportteja 85/2010 Working Papers. http://urn.fi/urn:isbn:978-951-44-8209-0. Accessed Oct 2015

Bergvall-Kareborn B, Hoist M, Stahlbrost A (2009) Concept design with a living lab approach. In: Proceedings of the 42nd Hawaii International Conference on System Sciences

Boland R, Tenkasi R (1995) Perspective making and perspective taking in communities of knowing. Organ Sci 6(6):350–372

Boronowsky M, Herzog O, Knackfub P, Lawo M (2006) Wearable computing Â¿ an approach for living labs. In: Applied Wearable Computing (IFAWC), 2006 3rd International Forum on, 1–8

Brasil D (2010) Experimenting with the urban experience: Rio, Lisbon and Weimar. A (re)search for creative collaborations and active exercises of citizenship. Dissertation of the der Bauhaus-Universität Weimar. https://www.uni-weimar.de/architektur/raum/doktoranden/thesesD_Brasil.pdf. Accessed Oct 2015

Campbell J (1997) Mechanisms of evolutionary change in economic governance: interaction, interpretation and bricolage. In: Magnusson L, Ottosson J (eds) Evolutionary economics and path dependence. Edward Elgar Publishing, Cheltenham

Celino A, Concilio G (2010) Participation in environmental spatial planning: structuring-scenario to manage knowledge in action. Futures 42(7):733–742

Chesbrough H (2003) Open innovation: the new imperative for creating and profiting from technology. Harvard Business School Press, Boston

Clark DN (1997) Strategic management tool usage: a comparative study. Strateg Chang 6(7): 417–427

Concilio G, Celino A (2012) Learning and innovation in living lab environments. In: Schiuma G, Spender JC, Yigitcanlar T (eds) Knowledge, innovation and sustainability. Integrating micro & macro perspectives, Proceedings of the IFKAD 2012 Conference

Concilio G, Molinari F (2014) Urban living labs: learning environments for collective behavioural change. In: Proceedings of the IFKAD 2014 Conference

Concilio G, Molinari F (2015) Place-based innovation: analyzing the "social streets" phenomenon. In: Proceedings of the IFKAD 2015 Conference
Concilio G, Moro A (2015) Trading zones and public spaces transformations. The case of piazza Leonardo da Vinci in Milan. Paper presented at the International Conference Group Decision and Negotiation, Warsaw, June
Cook SD, Brown JS (1999) Bridging epistemologies: the generative dance between organizational knowledge and organizational knowing. Organ Sci 10(4):381–400
de la Pena DS (2013) Experiments in participatory urbanism: reform and autogestión as emerging forms of urban activism in Barcelona. UC Berkeley Electronic Theses and Dissertations. http://escholarship.org/uc/item/1x0646mf. Accessed Oct 2015
Dougherty D (1992) A practice-centered model of organizational renewal through product innovation. Strateg Manag J 13:77–92
Dougherty D (2004) Organizing practices in services: capturing practice-based knowledge for innovation. Strateg Organ 2(1):35–64
Dutilleul B, Birrer FA, Mensink W (2010) Unpacking European living labs: analysing innovation's social dimensions. Cent Eur J Publ Policy 4(1):60–85
Ellström PE (2010) Practice-based innovation: a learning perspective. J Work Learn 22(1/2):27–40
Eriksson M, Niitamo VP, Kulkki S (2005) State-of-the-art in utilizing living labs approach to user-centric ICT innovation – a European approach. Center for Distance-Spanning Technology, Lulea University of Technology, Sweden. Nokia Oy, Centre for Knowledge and Innovation Research at Helsinki School of Economics, Finland. http://www.vinnova.se/upload/dokument/Verksamhet/TITA/Stateoftheart_LivingLabs_Eriksson2005.pdf. Accessed Oct 2015
Etzkowitz H (1998) The norms of entrepreneurial science – cognitive effects of the new university-industry linkages. Res Policy 27(8):823–833
Fischer G (2005) Distances and diversity: sources for social creativity. Proceedings of the 5th conference on Creativity & cognition. ACM, New York, pp 128–136
Følstad A (2008) Living labs for innovation and development of information and communication technology: a literature review. Electr J Virtual Organ Netw 10, Special Issue on Living Labs. http://www.ejov.org. Accessed Sept 2013
Forgues D, Koskela L, Lejeune A (2009) Information technology as boundary object for transformational learning. J Inf Technol Constr 14:48–58
Gibson H (2014) JPI Urban Europe Workshop Living Labs. http://jpi-urbaneurope.eu/jpi-urban-europe-workshop-living-labs/. Accessed Oct 2015
Harvey D (2003) The right to the city. Int J Urban Reg Res 27(4):939–941
Hou J (ed) (2010) Insurgent public space. Guerrilla urbanism and the remaking of contemporary cities. Routledge, Abingdon
Itami H (1987) Mobilizing invisible assets. Harvard Business School Press, Boston
Johansson LO, Lund Snis U, Svensson L (2011) Exploring brokering situations in an innovation boundary context. In: Selected papers of the Information Systems Research Seminar in Scandinavia, IRIS 34 – ICT of Culture – Culture of ICT, Trondheim, Tapir Akademisk Forlag
Juujärvi S, Pesso K (2013) Actor roles in an urban living lab: what can we learn from Suurpelto, Finland? Technol Innov Manag Rev, November
Kilpatrick S, Willis K, Lewis S (2014) Community action in Australian farming and fishing communities. In: Gallent N, Ciaffi D (eds) Community action and planning. Contexts, drivers and outcomes. Policy Press, Bristol
Leminen S, Westerlund M, Nyström AG (2012) Living labs as open-innovation networks. Technol Innov Manag Rev, September:6–11
Leonard-Barton D (1995) Well-springs of knowledge: building and sustaining the sources of innovation. Harvard Business School Press, Boston
Mäntysalo R, Balducci A, Kagasoja J (2011) Planning as agonistic communication in a trading zone: re-examining Lindblom's partisan mutual adjustment. Plan Theory 10(3):257–272
Marsh J (2008) Living Labs and territorial innovation. In: Cunningham P, Cunningham M (eds) Collaboration and the knowledge economy: issues, applications, case studies. Ios Press, Amsterdam

Michalski M (2006) Boundary objects and organizational integration. In: OLKC Conference, Coventry, 20–22 March. http://www2.warwick.ac.uk/fac/soc/wbs/conf/olkc/archive/olkc1/papers/232_michalski.pdf. Accessed Oct 2015
Milanovic F (2006) Travail d'articulation et organisations-frontières dans la recherché urbaine française. Soc Sci Inf 45(1):109–138
Miles MB (1964) Innovation in education. Bureau of Publications, Teachers College, Columbia University, New York
Molinari F (2011) Living labs as multi-stakeholder platforms for the eGovernance of innovation. In: Proceedings of the ICEGOV11 Conference, Tallin (Estonia)
Moroni S (2015) Libertà e Innovazione nella città sostenibile. Ridurre lo spreco di energie umane. Carocci editore, Roma
Nonaka I, Toyama R, Konno N (2000) SECI, ba and leadership: a unified model of dynamic knowledge creation. Long Range Plann 33(1):5–34
Pallot M (2009) Engaging users into research and innovation: the living lab approach as a user centred open innovation ecosystem. http://www.cwe-projects.eu/pub/bscw.cgi/1760838?id=715404_1760838. Accessed Sept 2013
Pettigrew A, Fenton E (2000) Complexities and dualities in innovative forms of organizing. In: Pettigrew A, Fenton E (eds) The innovating organization. SAGE Publications, London, pp 279–301
Sahlin-Andersen K, Söderholm A (eds) (2002) Beyond project management. Liber, Copenhagen
Sevón G (1996) Organizational imitation in identity transformation. In: Czarniawaska B, Sevón G (eds) Translating organizational change. W.de Gruyter, Berlin, pp 49–67
Spee AP, Jearzabkowski P (2009) Strategy tools as boundary objects. Strateg Organ 7(2):223–232
Star SL (1990) The structure of ill-structured solutions: boundary objects and heterogeneous distributed problem solving. In: Gasser L, Huhns MN (eds) Distributed artificial intelligence, vol 2. Morgan Kaufmann, San Mateo, pp 37–54
Star SL, Griesemer JR (1989) Institutional ecology, 'translations' and boundary objects: amateurs and professionals in Berkeley's Museum of Vertebrate Zoology, 1907–39. Soc Stud Sci 19: 387–420
Stenfors S, Tanner L, Hapalinna I (2004) Executive use of strategic tools: building shared understating through boundary objects. Front E-Business Res:635–645
Svensson J, Ihlström Eriksson C, Ebbesson E (2010) User contribution in innovation processes – reflections from a living lab perspective. In: Proceedings of HICSS'43, Kauai, Hawaii, 5–8 January
Sydow J, Lindkvist L, De Filippi R (2004) Project-based organizations, embeddedness and repositories of knowledge: editorial. Organ Stud 25(9):1475–1489
Trompette P, Vinck D (2009) Revisiting the notion of boundary object. Rev Anthropol Connaissances 3(1):3–25
Van den Broeck J (2008) Planning: a transformative activity. In: 44th ISOCARP Congress 2008. http://www.isocarp.net/Data/case_studies/1658.pdf. Accessed Oct 2015
von Hippel E (1986) Lead users: a source of novel product concepts. Manag Sci 32(7):791–805
von Hippel E (2005) Democratizing innovation. MIT Press, Cambridge. http://ssrn.com/abstract=712763. Accessed Oct 2005
Wallin S (2015) Living lab approach in urban development and planning – conceptual typologies of urban living labs. Presented at Workshop – Systemic Architectures for Sustainable Urban Innovation, Tampere and Espoo, 5–6 February
Wallin S, Horelli L (2010) Methodology of a user-sensitive service design within urban planning. Environ Plan B 37(5):775–791
Wallin S, Horelli L (2012) Playing with the glocal through participatory e-planning. J Commun Inf 8(3). http://www.ci-journal.net/index.php/ciej/article/view/883. Accessed Oct 2015
Walravens N (2013) The city as a service platform: a typology of city platform roles in mobile service provision. In: Proceedings of the Nineteenth Americas Conference on Information Systems, Chicago, IL, 15–17 August. http://www.ginevraresearch.com/images/freepaper/141_The%20City%20as%20a%20Service%20Platform_%20A%20Typology%20of%20City%20Platform%20Roles.pdf. Accessed Oct 2015

Warner ME (2011) Club goods and local government. J Am Plann Assoc 77(2):155–166
Worrall A (2010) Boundary object theory: concepts, propositions, and limitations. Working paper. Florida State University. http://www.adamworrall.org/portfolio/courses/lis6278/6278_paper3_boundary_object_theory_analysis.pdf. Accessed Oct 2015
Zdunczyk K (2006) Human boundary objects – fact or fiction? In: OLKC 2006 Conference, University of Warwick, Coventry, 20–22 March

Chapter 3
Smart Cities + Smart Citizens = Civic Intelligence?

Douglas Schuler

Abstract The title of this chapter is an algebraic way of introducing the relationship of three broad concepts which naturally will not succumb easily to simple relationships that "+" and "=" imply. The question mark at the end suggests that the relationship might not always hold. When faced with algebra problems, the basic goal is to isolate the unknown variable, which for the purpose of this exploration will be "smart citizens." The rules of the algebraic game always allow us to subtract equal quantities from both sides of the equation. Thus we can subtract "smart cities" from both sides of the equation in the title, which results in a new view of the same equation: Smart Citizens = Civic intelligence – Smart Cities? The purpose of this chapter then is to use the concepts of civic intelligence and smart cities to determine the value of smart citizens. It is this "value" of smart citizens that we must learn and then – moving out of the algebraic realm into the real world – contemplate how we ensure that we have sufficiently smart citizens to make up for what the smart city approach is unlikely to provide in our pursuit of civic intelligence.

Keywords Civic intelligence • Smart citizens • Smart cities

3.1 Civic Intelligence

From age-old problems like inequality, oppression, and natural disasters to new ones like bioterrorism, resource depletion, mass surveillance, nuclear annihilation, and climate change, problems seem to be growing faster than their solutions. If this is indeed the case, then our current approaches to governance are clearly insufficient. That, of course, seems to be at the root of many of the protests, citizen outcry, and overall dissatisfaction with governments around the world. In addition to lack of resources, incompetence, and corruption, governments may not have the flexibility, adaptability, and speed to adequately address today's problems. Things seem to be

---

D. Schuler (✉)
The Evergreen State College, Olympia, WA, USA
e-mail: douglas@publicsphereproject.org

G. Concilio, F. Rizzo (eds.), *Human Smart Cities*, Urban and Landscape Perspectives 19,
DOI 10.1007/978-3-319-33024-2_3

changing too quickly for the processes that we have historically relied on for them to be effective as quickly as they are needed. One reasonable approach, in theory – the one that is explored here – would be for broadly defined governance to be more widely distributed into the general population.

Of course, at some level the idea that people should play a strong role in their own affairs is noncontroversial. This rhetorical perspective is quite common yet not always earnestly pursued. For this idea to take hold renegotiation on many fronts between multiple sectors, between citizens and business, and citizens and government, the powerful and the powerless would be required. Business as usual would not be an option.

> Know the enemy and know yourself; in a hundred battles you will never be in peril.
> – Sun Tzu

The world, particularly the problems we face and the social web we now inhabit, is becoming more and more complex, and we, as a society, must become more complex in our thoughts and actions in relation to it. Troublesome situations are often the result of a myriad of factors, many of which are hidden. These situations do not have a single solution, will not be solved all in one time or once and for all, nor can they be solved by one person or a small group acting on their own. One does not unravel twenty-first-century problems with obsolete blunt instruments, nor even with trivial uses of sophisticated systems, clicking "like" on Facebook, for example (see Morozov 2013). Sun Tzu's adage above clearly captures this point: if we do not know the enemy – the problems we collectively face – and we do not know ourselves – our skills and our limitations – we have put ourselves in peril.

3.1.2 Will We Be Smart Enough, Soon Enough?

Time is critical. If we do not have the civic intelligence that we need soon enough, we will all face dire consequences and, of course, some will pay more dearly than others. This does not mean we need to run screaming down the street, even if it might be a perfectly natural reaction. Although that reaction may spur more people to pay more attention, it is not really likely to help clean up the mess we have created for ourselves. Our time would be better spent trying to figure out how to improve the coordination of our collective ability to address our problems. Every day that we delay in this task makes the task of addressing our problems more difficult.

While it may be true that tomorrow we will know more and/or be better suited for implementing the good ideas that we do have, those possibilities do not preclude the necessity of acting today. The consequences of our bad decisions (including not acting on good ones forcibly enough) are already in motion. Clearly the impetus is not just for doing anything, but for moving forward instead, with the aim of addressing our problems and the aim of improving our ability to address our problems.

3.1.3 Civic Intelligence Is a Good Name for the Unidentified, Yet Critical, Resource

Whether our problems become more manageable will depend on the ability of people to solve these problems equitably and effectively by working together. We do not presently have a concept that is widely recognizable for describing this ability. The absence of this important concept makes it difficult to talk about it with others and, hence, to consider it as a critical cultural resource. I use the expression "civic intelligence" to describe the ability of a society (or even a small group) to comprehend the problems that they are faced with and to develop equitable and effective approaches toward solving the problem (Schuler 2001). Civic intelligence is an expression that has been used sporadically for over a hundred years in ways that were generally compatible with our use. Our hope is that by popularizing the concept, it could help to inform and mobilize collective action.

One half of the idea of effectively addressing problems means efficiently making the consequences of the problems less severe, while diminishing the power of the processes that maintain the problem. But effectiveness does not tell the whole story. The word "effective" can assume a variety of meanings. Unfortunately, effectiveness is often characterized by a sort of efficiency, which can be evaluated mathematically in terms favorable to economic and political elites. That is why "equitably" is also key to the definition. It highlights the fact that the "solutions" to our problems are not really solutions at all if they are not equitable to the people (and the land) who are directly imperiled by the problems. Moreover, the types of problems we face (especially the ones that are brought on by presumably inherent human traits like jealousy, cupidity, or hatred) are not problems of logic or mathematics where typically there is an answer, and, once the answer is found, the problem is solved. If we solve the problem of finding the result of multiplying two numbers together, that problem is solved forever. Time will not erode the answer nor will the people revolt (or vote) and establish a different answer. We do, however, suspect that it is possible to reduce the damage from our inherent (though somewhat ill-adapted for current needs) baser instincts that are supported and magnified through a variety of social processes, institutions, and ideologies.

The idea of civic intelligence suggests that everybody has a role to play in the process of addressing shared problems. Civic intelligence needs to incorporate research to a large degree, but it has to be socially directed and followed through with action. This acknowledges that the research agenda is something that has to be negotiated and is not something that one needs credentials and an academic position work on. In all of our everyday lives, we maintain implicit hypotheses of what we believe to be true and which we are informally testing. In addition to the massive impact on the economic sector, business activities need special scrutiny. Business needs are of course to be considered within the civic intelligence perspective, but they must not be the sole determinant in terms of either provider or beneficiary of research.

Civic intelligence is applied to groups of people because it is through their interactions that public opinion is formed, decisions are made (or at least influenced), and actions are taken. It applies to groups, formal or informal, which are working toward civic goals such as environmental amelioration or nonviolence among people. This perspective is related to many other concepts that are currently receiving attention (to a large degree, because of the actuality and potential unearthed by new information and communication systems) including collective intelligence, distributed intelligence, participatory democracy, emergence, new social movements, collaborative problem-solving, human smart cities, social learning, and Web 2.0.

Intelligence is systematic: it uses what is known and innovates for new situations; it requires both thought and action; it acknowledges that we must push forward even when we do not know everything. And although we never know everything about anything, we do know quite a bit about some things and we can also use our awareness of our ignorance to be more intelligent.

If civic intelligence did not exist, it would be necessary to invent it. Fortunately, it exists to some degree in all individuals and groups and, at the same time, it could always be stronger.

3.1.4 Civic Intelligence: An Example

The image and text in Fig. 3.1 are from Liberating Voices (Schuler 2008), which contains 136 patterns for promoting civic intelligence. Each pattern acts as a seed that different people and groups will use in different ways. The Activist Road Trip pattern, for example, can be something as simple and unthreatening as taking a tour of local, social service agencies or as dangerous and historically significant as the efforts during Freedom Summer in the Southeastern United States to help register African Americans to vote in the 1960s. Our first pattern, Civic Intelligence, is both an example of an overarching topic and one "pattern" among many that is designed to help prompt people to be more engaged in civic research and action. The Civic Intelligence pattern can be used to spur just about any type of project, from organizing a neighborhood event to conducting a multinational campaign and everything in between and beyond. The sole constraint is that it must help equitably and effectively address a shared problem collaboratively.

The graphic associated with this pattern is a section of a mural that's in a low-income neighborhood in New York City. The mural illustrates the risks that are associated with asthma and shows all of the main elements of a repeating cycle, including causes of the disease, symptoms, testing, treating, and possibly dying from the disease. It also presents a doctor and the healthcare system. One does not need to be able to read to understand the story, and the images in the mural are all drawn from the local community. As such, it is an excellent example of homegrown civic intelligence. The case illustrated by the mural is also reflected by several other patterns as well, including Citizen Science, Power Research, Tactical Media, and The Power of Story.

Civic Intelligence

Civic Intelligence describes how well groups of people address civic ends through civic means. It asks the critical question: Is the society smart enough to meet the challenge it faces? Civic intelligence requires learning and teaching. It also requires meta-cognition — thinking about and actually improving how we think and work together.

Fig. 3.1 Image and text from the Civic Intelligence pattern card

3.1.5 Capabilities of Civic Intelligence

Based on our experience and research exploring the idea of civic intelligence for over a decade, we developed a framework that depicts the basic capabilities of civic intelligence. These capabilities have been suggested by a variety of studies from the perspective of various disciplines, each of which illustrate parts of the broader scope. The image below (Fig. 3.2) illustrates the main categories of these capabilities, which are discussed in more detail in Schuler (2014), from which the following summary is drawn:

- Knowledge: includes a variety of knowledge-based capacities such as theory, knowledge of problems, skills, resources, self-knowledge, and metacognition (thinking about and improving one's own thinking).
- Attitude and aspiration: include a variety of capacities that are typically seen as noncognitive but are essential for civic intelligence such as values, civic purpose, and self-efficacy.

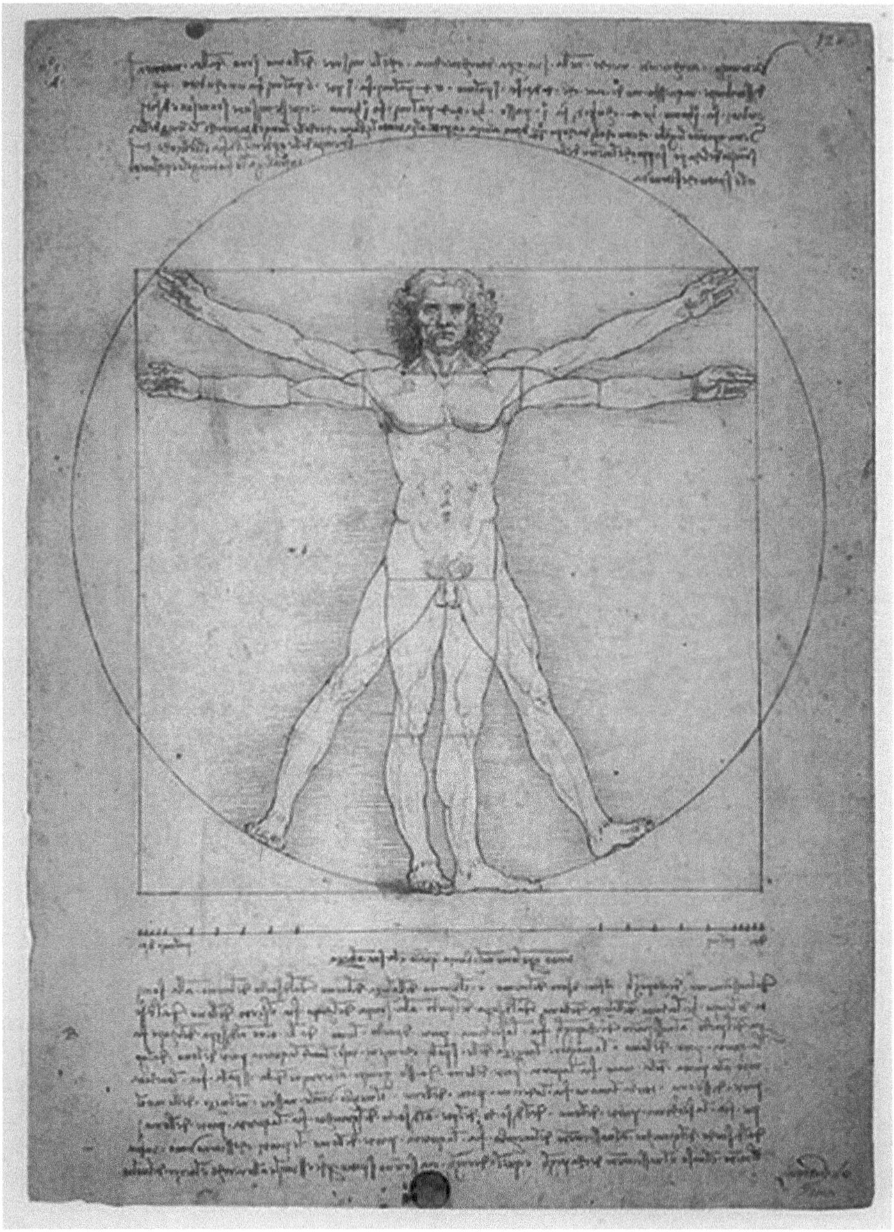

Fig. 3.2 In December 2013, Alicia Capp, a student in the Social Innovation and Civic Intelligence program at The Evergreen State College, depicted the five main categories of civic intelligence capacities in terms of interacting systems in a human form. She mapped knowledge to the head, attitudes and aspirations to the heart, relational capital to the outstretched arms, organizational capital to the core, and financial and material resources to the foundation we stand on. I used da Vinci's famous Vitruvian Man sketch, which is in the public domain, to demonstrate Alicia's characterization

- Organizational capital: includes the processes and structure of the collectivity that are needed to complete tasks effectively, such as personnel, work practices, and access to resources.
- Relational capital: includes reputation, social networks, trust, and opportunities.
- Financial and material resources: include money, buildings, land, etc.

People need to have knowledge, which has been (more or less tacitly) established as the sole purpose of current institutionalized educational systems. But without certain types of attitude and aspirations, the necessary pro-social focus may be lacking – and without organizational and relational capital, our ideas cannot be turned into something real.

While the use of these categories can be useful in the analysis of historical civically (or non-civically) intelligent activities, it is important to note that it is only through the dynamic and interdependent employment of the capacities that any civically intelligent action is actually enacted. The capabilities are used together. One use of the framework would be in planning, while another could be used for diagnosing or assessing a given group or organization for its ability to use and apply civic intelligence.

The framework encourages us to think about intelligence in a broad way. People who work with seniors talk about the importance of self-efficacy, for example, the idea that people need to feel that they have some control over their lives, that they have agency, and that they are capable of doing things, e.g., successfully completing tasks. This is a very important part of civic intelligence, which seems to be lost when people are focusing solely on technology. If people believe that they are incapable (especially if society also encourages this feeling of impotency), they are unlikely to start anything new. There are also some strong cultural memes that maintain social apathy: "That's boring! It's not fun to think about an ecological or a social problem." Worse, it is not "cool." Although there are millions of exceptions, the stereotype of youth is that the right clothes, music, and digital devices constitute the basic markers of coolness. However, cultural slavishness is not only an attribute of the youth. To buck the ideologies and folkways of one's peer community takes an exceptional person. William Gates (the father of multibillionaire Bill Gates) seems to be genuinely interested in building an equitable society at the same time that many members of his peer group seem to be obsessed with amassing more fortune. Unfortunately they are also successful in establishing ideologies that privilege wealth and the wealthy (and the policies to back them up), that power is something that can be purchased and the rich worked hard and deserve what they have and more, while the poor, through laziness or other inherent failings, deserve whatever misfortune comes their way (Piff 2013).

3.1.6 Civic Ignorance Interlude

Any study or deployment of civic intelligence necessarily introduces a counter-force at the same time: civic ignorance. Civic ignorance, the collection of forces that discourage and degrade civic intelligence, is always present but not always considered. Neither civic intelligence nor civic ignorance is a precise factor that can be characterized with a number. We need to avoid believing that we can be certain about either (which would be a display of civic ignorance!). Civic ignorance and civic intelligence are inextricable aspects of human existence. At the same time, it is important to work toward improving our civic intelligence, discouraging civic ignorance, and dampening its effects. Like civic intelligence, civic ignorance comes in various forms (Proctor and Schiebinger 2008). Some amount of ignorance is inevitable and even natural – we simply do not know everything and we never will. We forget information, make errors in reasoning, and disregard ideas that do not conform to our beliefs. Some civic ignorance, however, seems to be more conscious, calculating, and nefarious. This includes what could be called "the professionalization of ignorance," when vast amounts of time and money are expended on campaigns in which public ignorance is the sole objective. The campaign in the United States to downplay the health hazards of cigarettes, the deadliest artifact in history, is an important example, especially since many of the confidential documents that the cigarette companies produced have been made public. The current campaign to cover up and disparage scientific findings in relation to climate change (Oreskes and Conway 2010) is an important example of this because of its absolute relevance today in terms of the magnitude of the damage it can potentially cause. At the same time, it is critical to note that civic ignorance is being perpetrated – consciously and subconsciously – every day and at all levels. It is absolutely critical to note that acknowledging civic ignorance, even in ourselves, and endeavoring to understand it further is, somewhat ironically, key to the development of improved civic intelligence.

3.2 Smart Cities

While centered in Western Europe, the Smart Cities Movement is a worldwide phenomenon. It presents an innovative and important theme in relation to the infrastructure of cities. It generally means using "smarter" approaches, basically through the use of computers and smart data accumulation, integration, and analysis. The idea is to use data and data processing to get more efficiency, less waste and pollution out of utilities such as electricity, road systems, water, etc. While very few people would disagree with those aims, there are significant questions that need to be raised in relation to this movement. These will be taken up in the next section, but for now it should be mentioned that this is clearly a technologically focused argument. Moreover, it is most strenuously advanced by technologically focused corporations, whose primary mission is selling hardware, software, and technology-oriented services. The definition below from the Smart Cities Readiness Guide was developed by the SmartCitiesCouncil (2013), whose lead partners include AT&T, Bechtel, CISCO, GE Digital Energy, IBM, and Microsoft.

"A smart city uses information and communications technology (ICT) to enhance its livability, workability and sustainability. In simplest terms, there are three parts to that job: collecting, communicating and 'crunching.' First, a smart city collects information about itself through sensors, other devices and existing systems. Next, it communicates that data using wired or wireless networks. Third, it 'crunches' (analyzes) that data to understand what's happening now and what's likely to happen next."

We are reminded of the apocryphal young boy with the hammer who sees "nails" (or something that needs hammering) everywhere he looks. To a computer company, every problem that we face is crying out for a computer program to solve it. Note that "collecting, communicating, and crunching" would also be part of the civic intelligence perspective. After all, using a civic intelligence requires that people perceive ("collecting") problems, talk about them ("communicating"), and interpret information ("crunching"), but these chores cannot be accomplished solely with software, no matter how smart it may be. The complex and nuanced process of interpreting information should not be downgraded to the mechanical act of "crunching."

The Smart Cities Readiness Guide offers other definitions of smart cities. A US Office of Scientific and Technical Information report states that "a city that monitors and integrates conditions of all of its critical infrastructures – including roads, bridges, tunnels, rails, subways, airports, seaports, communications, water, power, even major buildings – can better optimize its resources, plan its preventive maintenance activities and monitor security aspects while maximizing services to its citizens." While mentioning the idea of "maximizing services," the predominant focus seems to be on monitoring physical structures and systems. Forrester Research goes beyond that with the assertion that "smart computing technologies" will make city services such as "city administration, education, healthcare, public safety, real estate, transportation and utilities more intelligent, interconnected and efficient." While computers can undoubtedly be put to good use in those areas, the implication that increased reliance on technology will yield greater and greater benefit is dishonest, whether or not it is intentionally implied by the smart city proponents.

The guide does acknowledge roles for people through three "supporting practices" that help ensure that the technology is "planned, deployed, and managed correctly." The first is policy and leadership and the second is finance and procurement. It is the third one, citizen engagement, that factors most directly into the topics of this chapter. Within the citizen engagement supporting practice, there are three points: "Continuously pursue two-way communication with citizens on deployment; offer an integrated, personalized citizen portal for services; and disseminate timely information about public safety, public health, transportation, and other services that impact the public." While mentioning citizen engagement is promising, the actuality is anemic. It implicitly asserts the view of citizens as consumers or customers, not as citizens. It reinforces the idea that experts will save us.

3.2.1 *What's Wrong with "Smart Cities?"*

Certainly elements of smart city plans are thoughtful and creative and will help us address many of our urban issues, especially the ones that can be more or less reduced to technical problems. The main problem with the smart city orientation is that it is an incomplete solution masquerading as a complete one. It can divert our attention in two basic ways. Firstly, if the smart city approach is seen as a total solution, it is more likely to get the lion's share of the resources. And why should it not? If it solved everything it says it will solve, then it is probably worth a lot. The second implication is that it pulls attention away from deeper problems that are likely to grow worse while we focus on technology. Both implications can lead to a decrease in the resources (of economic and human capital) available for use.

It also asserts that social issues will go away if, for example, it becomes easier for us to pay bills online. Left unchallenged, the smart city orientation is a form of technological determinism. It is what Evgeny Morozov calls "solutionism" (2014), the idea that a complex problem can be entirely solved, presumably forever, often with a simple remedy and usually through some mechanistic and non-negotiable approach. Arguments that offer total solutions are arrogant and dishonest. My main objection, however, is that they are stultifying – they cut off discussion where discussion matters most.

It is true that some, if not many, smart city texts mention the important role of people. Often, however, it seems like a second thought, which was added because someone had noticed that it had been omitted. Even when the text is present, the parts about citizen engagement are not always incorporated by the governments that purchase the advice and technological development from the smart city vendors. The smart city expectation that "data" is sufficient for city infrastructure management ignores the fact that social science "data" is not absolute, non-negotiable, without politics, or unequivocally and discreetly translatable into something else. The bad news for those who would prefer a clear and unambiguous world is that it is not true that we know precisely what social science data truly "means." It becomes dangerous to assume that complex social situations can be addressed through data crunching. Bill Gates is guilty of this when he claims that good (or bad) teaching can be strictly demonstrated via empirical testing. His belief that "bad teachers" are responsible for all inequality and could and should be used to drive educational policy is of the worst kind – especially if technocratic enterprises successfully sidestep discussion within the public arena. Calling education a type of infrastructure runs the risk of invoking this type of thinking (I am wary of this when I use the "civic infrastructure" expression, which I actually do in the next section).

How would we characterize the type of smart citizens we need to make up for the smart city deficiency in our equation? We believe that this type of person is more likely to emerge when a stronger democracy (Barber 1984) is called for, and people and groups feel that they are more valued and have a larger stake in the future.

3.3 Smart Citizens

Cities have the capability of providing something for everybody, only because and only when, they are created by everybody. – Jane Jacobs, *The Death and Life of Great American Cities*

It is not the intent of this chapter to bury the idea of smart cities, even if it could be done. The hope is to help challenge and add information to the discussion in a vigorous and somewhat radical way. This discussion ultimately must ask the question: what is a citizen? If the citizen is seen solely as a customer or consumer, then the government's job means providing services efficiently at a low cost (although we know that "efficiency" in the educational or social services sector takes different forms when seen from different perspectives. A person who lives in a gated upper class community and the recipient of government services to meet basic needs are not likely to see eye to eye on, for example, what is benefit and what is waste). Moreover, if the citizen is seen only as a periodic (or sporadic) voter who ratifies decisions merely by not paying attention, then the government may be swayed by other voices and apathy and ignorance become a habit among citizens.

We know groups have some civic intelligence, but some groups have more civic intelligence than others. Although there is apparently a sort of inertia at play, the breadth and depth of civic intelligence do not remain constant. One of the most important things about civic intelligence is that although the path cannot be always charted exactly, it can be improved. One obvious approach is to actually ask people to describe the problems they face. The next step would be to work with them to envision ways to help address the problems. Note that computer technology could – and probably should – be used to support this approach, but computer technology is not essential to this, and the smart city imperative is not likely to prioritize this difficult-to-quantify route.

Not focusing on humans can also lead to less usable and effective systems. It would mean missing the most interesting and important part, i.e., the relationship of people with technology, which affects their lives directly and indirectly. If the focus on technology succeeds in eclipsing the (ideally) complementary focus on the social element, it is quite possible that it will lead to solutions that are really neither as effective nor as equitable as they could be. The downside is that efforts and resources are put into a technological solution while ignoring the most important aspect, the human element. This is a pattern that we seem to repeat over and over again, a combination of commercial pressure, promises, and the decision makers' apparent inability to adequately view the richness of the situation. Perhaps it is necessary to talk about civic infrastructures, to emphasize both the smartness and the degree of engagement of the citizenry as critical aspects of an enlightened city, which cannot be provided through technology alone.

A smart city without smart people is impossible, but "smart" cannot just mean the ability to solve mathematical or logical puzzles quickly, since our problems are of a different sort entirely. The idea that the world now needs merely the existence, identification, or creation of smart people – those who test higher on IQ tests, have demonstrated ability to solve puzzles, or have skills in science, technology, engineering, or mathematics (STEM) – is widespread. It is not only up to them to hold the beacon aloft for the rest of us. Moreover, a significant part of our ability to address our big problems is social, because addressing real problems in a real way means addressing them collectively. Addressing climate change, to choose an obvious example, will not be "solved" by a small elite who will do it for us, even if made up entirely of geniuses. We clearly need "smart" people for our smart cities, but our hope is that their smartness is augmented with something that is closer to what has been traditionally called wisdom – and what we are calling here civic intelligence.

Smart citizens are not merely the tenders of the smart city apparatus. Every stage and focus of a smart city deployment must be scrutinized, but it does not end there. The entire enterprise must be examined not with the aim of stopping it, but to shape the effort, to bring critical omissions to the surface, and to intervene appropriately.

The significance of Human Smart City (Marsh, Molinari and Rizzo 2016) and other related efforts is that they aim to challenge and change a technocentric approach into one that is more centered on the human aspect and less motivated by profit. Adding "human" avoids the misleading view that technology "solves" problems by itself. It asks how we can set up technology platforms that help make urban social systems more equitable and effective at the same time. It is a long-term, incremental, participatory design process that integrates experimental, educational, and community mobilization, research, and policy work within a shared, coherent perspective.

3.3.1 Civic Intelligence in the Urban Environment

At some point in the last decade, a remarkable turning point was reached: for the first time, the majority of the earth's population lived in cities rather than in the countryside. With urbanization, we have seen many changes including the way we observe the city. The urban/rural divide has become more pronounced. This shift in relative populations has helped enable much broader shifts than had been anticipated, both qualitative and quantitative. For one thing, the new concentration of people has greatly increased the opportunity for idea circulation and economic innovation. Additionally, power is centralized in cities and although cities obtain food and vast amounts of resources from outside the city, there is a perception that cities do not rely on the actual (physical, mostly nonurban) environment. Civic intelligence is a general perspective that can be applied to any collectivity, but due to its enormous influence, the city deserves special attention.

Most decisions regarding finances and economic priorities are made in cities. Media production is centered in cities as is virtually all political wrangling. While resources from nonurban locations are absolutely essential, the conditions under which the extraction (of minerals, coal, trees, fish, food) is conducted are largely determined by people living in vastly different realities. While the first victims of this asymmetric relationship will likely be the land, the people, and the other living things that are close to the places where the needed resources are found, those living in cities, sheltered to some degree from events elsewhere, will ultimately feel the pain of ill-considered and abusive policies that were passed downstream.

3.3.2 *Maintenance of Society, Culture, and Civic Intelligence*

Institutions, both formal and informal, are needed to help perpetuate culture and societies. These can promote or retard civic intelligence. Societies and cultures are continuously being created and recreated by their institutions and their members. They can also be degraded when this recreation is interrupted. Interruptions occur periodically as a result of natural disasters, wars, poverty, epidemics, prolonged political struggles, mass displacement, or various addictions. The bottom line is that some critical mass of individual people in those societies must meet certain conditions, and the culture – and its civic intelligence – may falter or fail depending on them.

At the most basic level, people need to be safe and have the ability to earn a living. Beyond that, people need to feel that they belong in society and this will be accompanied by the feeling that society, at some level, is legitimate. This promotes a feeling of purpose and puts people in a position in which they are more likely to contribute ideas and other resources and, more broadly, to participate in the direction of society and culture. In other words, they become active, participating citizens. Building on this, it should be possible to provisionally identify indicators that reflect these basic needs and the level of civic intelligence in a given collectivity.

Although it is probably impossible to actually measure civic intelligence, there are many interesting directions to go in which lead to the development of meaningful, though not definitive, indicators. Even though civic intelligence is probably best evaluated with actual cases, by looking at cities, regions, and countries, it is hypothetically possible to gain an idea about how civically intelligent they were by using some of the indicators listed below:

- Knowledge of the environment, natural, and otherwise
- Social (political, educational, cultural) engagement
- Social capital
- Health and well-being
- Economic and other opportunities
- Relative equality of inhabitants
- Transparency/lack of corruption
- Good neighbor

At this point, we are not interested in boiling this down to one or two generalized values that seemingly account for the sum of civic intelligence in a city. In fact, some values may be hard or impossible to pin down in certain circumstances and others can stand in. For example, according to Stephen Bezruchka (2000), "Nothing affects the health of populations as much as the amount of social and economic hierarchy observed in that population. The greater the gap between the rich and the poor, the worse the health, as measured by mortality figures or by self-reported health."

Whether they seem that way or not, these indicators are measurable (albeit non-precisely) and are both the result of social processes and key engines (or enablers) of social processes. The indicators are interrelated and interactive, e.g., if there is a high level of social engagement and knowledge of the environment, it is more likely that problems are successfully addressed, hence more civic intelligence. Health is one of the important and main indicators, particularly in economically diverse communities, because it demonstrates that a perennial, essential, and universal issue is being addressed. The holistic/systemic nature of these indicators is reflected in an important article, "Owning the City" (de Lange and de Waal 2013): "To engage people with communally shared issues, it is essential that people envision themselves as part of the urban fabric and understand that their individual actions make a difference to the common good. They also need to trust other urbanites to act accordingly."

Finally, because no human settlement is autonomous, it is necessary to consider the good neighbor indicator, which focuses on interactions with external communities. For example, in wealthy neighborhoods in the United States, one could look at the other indicators (e.g., health and opportunities) and conclude that they must have a lot of civic intelligence. Unfortunately, it may be the case that they are exporting their problems to other neighboring cities that do not have as much money, or they are using more than their fair share of resources or barring (or even expelling) the less fortunate from their premises. Moreover, it could be the case that they are actively exploiting other regions. Having a low good neighbor indicator would remind communities with more resources (financial or expertise related) that they are not helping communities with fewer resources and they might be obligated to improve. Low numbers for this indicator would demonstrate diminished civic intelligence, possibly coupled with high levels of corruption. If there were no "good neighbor" requirement, a rich, gated community might inaccurately demonstrate the pinnacle of civic intelligence. Crime could be low, health factors would be high, and there would generally be little rancor to spoil the enjoyment of what money can buy, i.e., fine food, art, leisure, travel, and health. Furthermore, the side effects (such as unacknowledged privilege and a skewed sense of reality) could move the community away from any willingness to help address problems and turn it toward exporting the problems. This is one area in which sustainability measures and programs are directly relevant.

While these indicators can (we suspect) describe the state of civic intelligence in, say, a city, these indicators are also likely to be useful as a broad aspirational focus. In other words, they can be used to establish goals in addition to evaluating progress on meeting social objectives. Last but not the least, it is important to remember that the responsibility for improving the seven measures above – and improving civic intelligence in general – is not the responsibility of the government alone.

3.4 Smart Steps

3.4.1 We're All Laboratories Now! (or at Least We Should Be)

Fiorella De Cindio and Peraboni (2010) have asserted that Italy could be thought of as a laboratory, basically a location where experiments are conducted. Italy, a well-established democracy that recently "has suffered from 'democratic anomaly' where the prime minister is also the owner of a large media empire," is the location of ideas and projects under consideration and many other ongoing ones that are intended to address social ills. For them, the Internet provides a good portion of the raw medium upon which society experiments with new ways of organizing, challenging, collaborating, telling stories, and, in general, being active and engaged citizens.

The amount of experimentation may be high because Italian society has a culture of experimentation. It may also be because there is a shared perception that there is an urgent need for new approaches. Universities, civic groups, and institutions are all working in this field. But Italy is not the only laboratory – all countries have this feature to some degree and they assume different forms in different contexts. Some of these labs may succeed and some may fail. Everywhere, however, people need to be part of the solution. We cannot expect institutions to solve all our problems for us. That approach often fails. Authorities many pay little attention if there is no involvement or interest shown. We have to have our eyes on the future, not only on today.

Sassen (2011) reports that "Wherever I go in the world, I find at least some technologists, urbanists and artists who are beginning to 'urbanize' technology." She goes on to say that "When this happens, the city becomes a heuristic space; it talks with the average resident or visitor rather than simply commanding them." But the amount of experimentation and development of new policies, services, artwork, and computer applications is not determined solely by whether a city, region, country, or planet is a "lab" (nor can we simply begin calling our cities labs and congratulate ourselves as Monsieur Jourdain did, when he was struck by the insight that the words he had been uttering all his life were actually prose). De Lange and de Waal (2013) warn that artistic and other urban interventions often "remain highly temporary and stick to oppositional politics." They argue for an alternative approach to "urban design with digital technologies that focuses on the active role of citizens and uses the city itself as the test bed for experiments."

The basic necessity of the lab is consciousness of its purpose, activities, and strategies. Reflection is key. Generally, the activities of the lab are to be purposeful. This means that some possible future is desired and some criteria will be applied – at the end at least – to judge success. A "lab" also seeks (and creates) knowledge and employs resources to this end. This also means that criticizing the present state – although this is often implicit, i.e., the desired state is not the current state – is necessary if we are to move forward.

When we think of our society as a lab that is consciously launching experiments, we are reframing the world in our minds as, potentially, a vast experiment in progress. With this type of reframing, we would do fewer things only for private gain or with no regard to consequences. We would incorporate civic intelligence into a new type of "civic infrastructure" with citizen think tanks or civic intelligence labs.

3.4.2 Ideas and Recommendations

The direction or trajectory of technology should not be left to the technocrats. Computer companies should not run our cities, regardless of their vast resources and their can-do attitude. While not sacrificing the idea that efficiency (to mention just one benefit that is claimed) is important, we need to be exploring concepts that enlarge the scope and depth of civic necessity, potentiality, and possibility – in one word, civic intelligence. Rather than dead-end progress via the solutionism of software vendors, we need to be considering our urban future as an opportunity to consciously evolve, in which the vendors could be involved as well. Although some type of cultural clash may be inevitable (given some basic differences in perspectives), they could work in concert with activists and other citizens on applications and with our rural counterparts in a mutual learning process.

Unfortunately, people in the technological community are more likely to think in terms of solutions. At the same time, people in the non-tech community are often intimidated by techspeak, impressed by the confidence and assurance that often accompanies the technological sales approach. At the same time, people are eager for solutions to their problems. But the "solutions" of the non-tech variety are not viewed as "solutions" at all. The suggested approaches are not advanced with the certainty found elsewhere and they often only suggest vague directions. More conversation in plotting the way forward is needed and the kinds of conversations suggested here are not merely chitchat. They are purposefully reflective and civic action – sometimes small and sometimes vast – is always a possible outcome.

The primary job is to expand the conversation and to put these efforts on the public agenda. The new technology has allowed an amazing amount of new development as well as new opportunities and ideas. Unfortunately, both the intellectual inertia and the investment inertia are arguing against democratic engagement and control (to have an idea of the magnitude of the inertia, think back on the time from 1995, not two decades ago, when all commercial activity was forbidden on the Internet, to now, when commercial activity is the norm and the commercial entities seem to call all the shots). Promoting citizen engagement, including that around technological deployment, will not become part of the conversation without effort. City governments need to make commitments in advance to public involvement and oversight. Citizens should be placed on committees. Cities could make grants available to neighborhood groups with creative ideas, that want to become involved with new approaches through technology or non-technology-oriented projects involving the increased understanding of city systems. And city governments should sponsor public meetings with vendors at public locations like libraries.

The opportunities afforded by new technology, although overshadowed by commercial activity (largely commodity and entertainment based), have not been obliterated yet. However, in the era of "big data," in which highly educated technology experts devote their careers to analyzing people's online behavior on Facebook or Twitter, in order to sell more services and merchandise, it is easy to become distracted from what really matters.

One approach that today's problems seem to demand is transdisciplinarity. This embraces "close collaboration between researchers and community stakeholders who work together to understand and ultimately resolve societal problems" (Stokols 2014). According to Stokols, drawing from a variety of studies, "Cross-disciplinary teams have become increasingly prevalent across many research domains, owing to the growing recognition in academia and society at large that the world's most complex and intractable problems, – including global climate change, poverty, war, famine and disease– can be better understood and ameliorated from a broad interdisciplinary perspective than from the narrower vantage points of separate fields." In the same work, Stokols also states that the "scholars who possess diverse knowledge sets drawn from multiple fields, as well as the inclination to integrate multiple analytic levels in their work, are more likely to generate highly radical innovations as compared with those whose knowledge and conceptual strategies are more narrowly circumscribed." Although these new collaborations are not trivial to institute or sustain, they are vital. It will be important to work directly with mediators, i.e., civil society organizations and the media, but also to work directly with artists, educators, designers, community health workers, social workers, business owners, activists, and marginalized communities.

Transparency of information is a good antidote to possible excesses of government and business, including the future deployment of smart city ideas and systems. This means access to the communications, agreements, transactions, and meetings, but also access to what the systems themselves produce, i.e., data on electricity, pollution, and street usage patterns, to suggest just a few.

While opening the discussion and introducing new ideas is crucial, it is also vitally important to consider the work of artists, researchers, educators, activists, community advocates, and other people who have been thinking about and working within this context for years. Some of the more intriguing ideas around technology, the city, and people include work in participatory design, in which people are engaged in the design of systems and artifacts that affect them. A variant of this is participatory sensing which, for example, could allow more people to investigate pollutants in their own neighborhoods.

Saskia Sassen (2011) and others have advanced the intriguing idea of making the new infrastructure visible. Although traditional infrastructure generally has a palpably visible form (pipes, wires, bridges, roads, etc.), the new digital-based infrastructure, which is increasingly conjoined with the old, could be made easier to understand. This could be done in a variety of ways, making it more visually or aurally intriguing, for example, as well as educational in the sense that the flows, switches, nodes, etc. of the new overlaid technology would reflect its actual use within the city (some of these ideas are reminiscent of Stafford Beers' proposed Cybersyn project in Chile in the 1980s, while many now are being carried out in cities around the world in various control centers or monitoring centers).

The idea of "community networks," whose popularity crested in the 1990s, still seems plausible – or at least desirable – today, but with caveats. The first is because information and communication technology have entirely changed since then. The second is because the functions that the new networks address should probably change. One approach would be to actually develop deliberative (and other collaborative) systems, which could ideally promote the collective building of things, including decisions and ideas. New wireless community networks are also being developed around the world. These projects are likely to face many of the same challenges as the earlier generation. Yet, arguably, the ability to launch new communities/networks that are transdisciplinary will be crucial if civil society is to successfully organize itself to promote civic intelligence, to engage with the problems we face, and to mount successful challenges to the powers that will be.

One intriguing idea hinted at earlier is reconciling urban lifestyle and urban perspective with deeper, meaningful relationships for the people and territory outside the city. What resources are we taking from outside and what impact does that have? In our increasingly urbanized world, we should continue to build collaborative networks of urban professionals and citizens but we should also welcome rural participants. We should evaluate policies that cities adopt and compare them with the rest of the world, to receive suggestions and avoid mistakes that other cities have already made. Examples can be identified in the decisions relating to forests (deforestation), to rivers (pollution), and many others. Reestablishing our urban/rural links and redesigning a less exploiting relationship is an important step for cities and for everybody else.

Many social problems are universal yet assume different forms depending on their context. Corruption is one such problem (Cockcroft 2013). It can be endemic in one location and absolutely paralyzing and be much reduced and almost "under control" in another. At any rate, rooting out corruption is both a local and a global activity in which citizens could play important (though potentially risky) roles working, for example, with each other and with media outlets and government offices where appropriate.

3.4.3 Challenges

The challenges we face are enormous, partly because of the two aspects involved. The first is that the world itself is beset with bewildering complex problems. The second is ourselves. To address these issues, we must change ourselves and our institutions. If this required an entirely different type of human being, we would be in even deeper trouble. It is my belief that individuals around the world would need to change their activities only slightly for more major change to occur. People are already working toward positive change, for the rights of others (including the natural world), and against oppression. They are already thinking about how their lives can affect others and what changes they could make in their own lives. They are already reaching across national and other divides, and competing groups are learning to transcend boundaries, to mend fences, and to work together. Now, if more people were to do this and they were a bit smarter about it – in other words, if they demonstrated increased civic intelligence – change on a broad scale would indeed be possible. Collaboration is crucial to this effort and how smart we are about how we collaborate is probably even more crucial. We need to collaborate, directly and indirectly. While metacognition is something that separates experts from nonexperts at the individual level, it is especially critical to consider this in relation to collaboration. The new technology can help with that but it will not do it for us.

Cities are not machines. But like machines, people use cities and breathe life into them. Humans must take care not to be enslaved by them or forget their own worth and influence. Cities are places that are built, inhabited, and modified by people. People animate cities, and cities – whether they are smart or not – are merely ghost towns or ruins in the absence of people.

We need smart cities. But without a vigorous, aware, ubiquitous, and diverse contingent of smart citizens, we will not develop the civic intelligence that is desperately needed. The equation in the title would not balance. To address the problems we are facing, we will need creativity, dedication, humor, reason, and compassion. Fortunately, the potential is there because people often have these talents.

Remember that governance is not solely a technological matter. The market or side effects will not solve our problems for us. But citizen engagement is not "one size fits all." Different contexts require different approaches. We cannot define a single, uniform, "silver bullet" that is guaranteed to work everywhere.

The informed contribution of citizens is an indispensable element of governance. Citizen engagement ideally provides both impetus for social change when it is needed and a bulwark against tyranny and oppression when that becomes necessary. These are absolutely key roles and their importance highlights the need for civic intelligence. Without strong, engaged citizenry, we will not be able to address our problems. With strong, engaged citizenry, we may be able to address our problems.

References

Barber B (1984) Strong democracy: participatory politics for a new age. University of California Press, Berkeley

Bezruchka S (2000) Are the rich making us sick? Scientific research shows that inequality leads to poor health. The Washington Free Press, July/August. http://wafreepress.org/46/free_thoughts.html. Accessed 24 July 2015

Cockcroft L (2013) Global corruption: money, power, and ethics in the modern world. University of Pennsylvania Press, Philadelphia

De Cindio F, Peraboni C (2010) Internet as a platform for political engagement: from protests to proposals. In: Proceeding of the CIRN – DIAC community informatics conference 2010

de Lange M, de Waal M (2013) Owning the city: new media and citizen engagement in urban design. First Monday 18(11). Retrieved May 08, 2016, from http://journals.uic.edu/ojs/index.php/fm/article/view/4954

Marsh J, Molinari F, Rizzo F (2016) Human smart cities: a new vision for redesigning urban community and citizen's life. In: Knowledge, information and creativity support systems: recent trends, advances and solutions. Springer International Publishing, Switzerland, pp 269–278

Morozov E (2013) To save everything, click here: the folly of technological solutionism. Public Affairs, New York

Oreskes N, Conway E (2010) Merchants of doubt: how a handful of scientists obscured the truth on issues from tobacco smoke to global warming. Bloomsbury Publishing, New York

Piff P (2013) Wealth and the inflated self class, entitlement, and narcissism. Personal Soc Psychol Bull 40(1):34–43

Proctor R, Schiebinger L (eds) (2008) Agnotology: the making and unmaking of ignorance. Stanford University Press, Redwood City

Sassen S (2011) Talking back to your intelligent city. McKinsey, 1 February

Schuler D (2001) Cultivating society's civic intelligence: patterns for a new 'World Brain'. J Soc Inf Commun 4(2):157–181

Schuler D (2008) Liberating voices: a pattern language for communication revolution. MIT Press, Cambridge, MA

Schuler D (2014) Pieces of civic intelligence: towards a capacities framework. E Learn Digit Media 11(5):518–529

Smart Cities Council (2013) Smart cities readiness guide. Smart Cities Council, Redmond

Stokols D (2014) Training the next generation of transdisciplinarians. In: O'Rourke M, Crowley S, Eigenbrode S, Wulfhorst J (eds) Enhancing communication and collaboration in interdisciplinary research. Oxford University Press, New York, pp 56–74

Part II
Complex Participatory Design Processes

Chapter 4
Design and the Transformation of Cities

Alessandro Deserti

Abstract The relation between design and the city has changed considerably over the last years. Quite a few factors have interacted to produce this change: some are bound to the evolution of design culture and practice itself; some are bound to the transformation of cities; and some are bound to the transformation of social relations and their interaction with technologies.

In this context, the traditional disciplines of urban planning and architecture have undertaken profound transformations, but in the view of the author, the major changes occurred in the design that used to be focused on the small scale, where we had a progressive expansion of the territories of interest and application, which completely changed the role that design can play in the transformation of cities.

The purpose of this chapter is to offer an overview of the evolution of the relationships between design and the city. The author discusses the main practices of design applied to cities from the early 1970s to the present day. The discussion introduces the different practices of design for cities as a consequence of the radical transformation of the design discipline. Then, evolution of some experiences, like that of the Milan "Fuorisalone", is illustrated as clear representations of how the visions of the city that the author described coexist and are connected with the steady evolution of the culture of design.

Keywords Design culture • Transformational design • Design context

4.1 Design and the City: Visions, Approaches and Practices

The relation between design and the city has changed considerably over the last years. Quite a few factors have interacted to produce this change: some are bound to the evolution of design culture and practice itself; some are bound to the transformation of cities; and some are bound to the transformation of social relations and their interaction with technologies.

A. Deserti (✉)
Department of Design, Politecnico di Milano, Via Durando 38/a, 20158 Milan, Italy
e-mail: alessandro.deserti@polimi.it

G. Concilio, F. Rizzo (eds.), *Human Smart Cities*, Urban and Landscape Perspectives 19,
DOI 10.1007/978-3-319-33024-2_4

The traditional subdivision of the design tasks was based on the scale of the objects to be designed, ranging from the whole territory or portions of the city (urban planning) to single buildings or structures (architectural design) to "accessories" such as urban furniture (product design). Within this subdivision, urban planners and architects were in charge of shaping the city: even if the idea of a cross-scale design ("from the spoon to the city", to say it with Ernesto Nathan Rogers) was expressed, specialisation has become the norm.

Today, the interaction between the material substrate constituting the city as we historically know it and the possibilities that the cyberspace is offering to reshape the relations between individuals, social groups, tangible objects and systems has led to a much more complex situation. Professional practices that used to be vertically specialised and focused on specific objectives are displaying many intersections and areas of overlap, and new practices are emerging.

Both urban planning and architecture have undertaken profound transformations, but in our view, the major changes occurred in the design that used to be focused on the small scale, where we had a progressive expansion of the territories of interest and application, which completely changed the role that design can play in the transformation of cities.

The first driver of this change is the shift from the concept of the city as a system of tangible artefacts (the whole city, the infrastructures, the buildings or the "accessories") to the concept of the city as a system of relations based on a tangible substrate. This is in truth a long-standing concept that has encountered many difficulties in being concretely applied. As Landry and Bianchini put it:

> In this century the main solutions went a step further, based on the theories of how to create 'the good city' associated with authors like Patrick Geddes, Lewis Mumford or Jane Jacobs. They emphasised not only how a city might be shaped physically but also what could improve people's lived experience of cities. Yet when these ideas were taken up by the emerging planning profession, they were interpreted mainly in physical terms, disregarding the more subtle psychological effects on people. (Landry and Bianchini 1995: 13)

Even if the idea of the city as a social system is far from being new, the recent transformations are connected to (and empowered by) the overall change of the economic paradigm, from an economy primarily based on the exchange of physical goods to an economy largely based on the exchange of services. The introduction of a service-dominant logic (Vargo and Lusch 2004) called for the capacity of giving shape to complex bundles of services, infrastructures and tangible goods and led to the development, experimentation and wide adoption of new design processes and tools. This transformation affected the design culture at large, but the city emerged as a paradigmatic field of application also because, due to the steady process of urbanisation, cities themselves have never been as successful as today and at the same time never as critical and challenging.

The relation between the tangible and intangible layers of the city, to which we refer today with the somehow abused term "smart city", constitutes the overall engine of the recent changes, but different pathways have characterised the transformation of the relationship between design and city. These pathways are bound to different visions of the city, which bring along specific approaches, forms of practice and roles of design:

1. The city as a product or brand
2. The city as a space for creativity
3. The city as a space for services
4. The city as a participated construct
5. The city as a complex system

These visions are at some extent in contradiction one with the other, but they can be represented along a continuum: all of them originate from specific ideas and are related to different historical and cultural moments, but they coexist and are interwoven.

A synthetic account of these visions will give the possibility to investigate the new roles of design in the transformation of cities.

4.1.1 The City as a Product or Brand

Starting from traditional consumer goods, the field of marketing undertakes a remarkable expansion of its territories of application, in which almost anything can now be seen as a product to be marketed. In this line of thinking, territories can be seen as offerings, competing in local or global markets, and cities represent a specific kind of product that can be positioned, communicated and marketed. Hence, most of the techniques applied to products in a competitive environment can be (and actually have been) applied to territories and to cities specifically. Going further, if we look at cities themselves as concentrated market spaces, single areas within the same city may be seen as competitors (Mäding 2006) or as actors that operate in a regime of "co-opetition" (Brandenburger and Nalebuff 1996).

The shift from the idea of the city as a product to that of the city as a brand represents a further step in the same line of thinking. Place branding emerged during the 1990s as a new paradigm in a growing global and local competition among territories (Kotler et al. 1993; Gold and Ward 1994), sometimes assuming the name of "destination branding" and being primarily focused on the touristic market (Morgan et al. 2002).

In the perspective of dematerialisation that we introduced, this passage is quite interesting, since the brand can be described as an intangible asset based on a tangible substrate. In this frame, the need of managing the interrelation between the tangible characteristics of places and the intangible nature of brands emerges as a key issue of place branding (Anholt 2007). At the same time, the limited possibilities of reshaping the "product" in accordance to the brand make place branding quite different from traditional branding practices, leading to an inversion of roles in which the product determines the brand, despite the overall strategic attitude of branding. This is one of the causes of the cosmetic attitude of many place branding initiatives.

Even if the competitive positioning of territories is a complex issue, bound to their long-term heritage and to long-term investments in their tangible and intangible assets (resources, forms of capital, infrastructures), the idea of "competitive identity" as expressed in place branding (Anholt 2007) has often led to operations where design is asked to give shape to the expression of territorial brands through the development of a visual identity (VI). As already noted, in the majority of cases, the approach is cosmetic: the city as a product or brand is assumed with its already existing characteristics, and the attention is primarily focused on building an appealing VI. The typical results of these operations are logos, taglines, communication campaigns, touristic routes with their websites, applications, brochures and signs. In other cases, even if the VI remains a fundamental aspect of place branding, a more profound work of interpretation and of strategic redirection of the brand through long-term operations has been done.

4.1.2 The City as a Space for Creativity

The concept of the creative city (Landry 1990, 2000; Landry and Bianchini 1995) represents a further step in the process of "dematerialisation" of the urban environment. In its more profound vision, it introduces the idea that the new pressing challenges that contemporary cities are rising can be faced only through a leap in the forms and the processes of governance, in which the imagination and creativity of urban actors and stakeholders help solve wicked problems.

The concept of the creative city is underpinned by the idea of an overall shift in the economic paradigm, where the interaction between creativity and culture taking place in urban environments can produce economic benefits (Landry 1990). Nevertheless, if we shift to practice, this wide concept seems to be overlapped and blurred with the transformation of districts where creative activities concentrate or are concentrated. Landry (2000) describes these places as "creative milieus", most often resulting from the conversion of downtown areas, where new communities are located. The combination of "hard" and "soft" infrastructures is the most relevant characteristic of creative milieus: within this frame, the city starts being described as the intersection or juxtaposition of a tangible layer, with its physical features and visual image and an intangible layer, made up of social relations, human interactions and flows of ideas.

To be objective, we should underline the pitfalls and the twofold nature of the transformations that have led to contemporary creative districts: on the one hand, we may see them as cases of revitalisation of deprived neighbourhoods and brownfields through the introduction of new activities in which creativity and culture play a major role; on the other hand, we must recognise that the introduction of these activities and of their related communities may be primarily meant to support real estate operations, in which the increased price of buildings leads to gentrification and social exclusion.

The location of art and design communities, in their expansion from the elite to the mass, has led to many of these twofold urban conversions. This process has been formatted and internationally applied with local variations to give shape to a variety of similar art, design and creative districts around the world. These cases may be more or less successful, but they somehow betray the original idea of the creative city, whose more interesting aspects should be found in the holistic and open perspective on the governance of the city and in the focus on "people's lived experience of cities".

4.1.3 *The City as a Space for Services*

Services are by far the main economic activity of contemporary cities: citizens themselves require a growing amount of services, and services in general tend to be concentrated in urban and metropolitan areas, where the service economy thrives. Seen from our perspective, services represent a perfect metaphor of the overlap of tangibles and intangibles. Services may be described as processes whose design often implies the concurrent design of tangible artefacts that support people's navigation along the same processes.

The sedimentation, in the last decade, of the concept of the smart city[1] has strongly influenced the recent success of service design as the approach that better fits the need of municipalities to face service innovation. In a period of crisis and a profound renewal of the welfare system, service design is gaining momentum as a methodology capable of supporting the implementation of public services as the result of a co-creation process. In this vision, the design and the delivery of services is a participatory activity that involves citizens, public bodies, businesses and third sector operators in complex forms of interaction, where the traditional distinction between providers and users becomes blurred.

The success of service design in the context of smart cities – and more in general its progressive entrance in the public sector – primarily relies on the suggestion to overturn the conception of services from the dominant paradigm that moves from technologies to solutions to the emerging one that moves from problems to solutions (design thinking approach). Service design, heavily rooted in the tradition of user-centred design – people centred, design led and based on projects – is expected

[1] At its core, the idea of smart city is rooted in the creation and connection of human capital, social capital and information and communication technology (ICT) infrastructures in order to generate greater and more sustainable economic development and a better quality of life (Directorate General for Internal Policies 2014). In truth, there are many perspectives on smart city: some focus on ICT as a driver and enabler, while broader definitions include socio-economic, governance and multi-stakeholder aspects, such as the use of social participation to enhance sustainability, quality of life and urban welfare.

to bring higher quality to users and providers, new business opportunities and new methods and tools to deal with innovation in the public sector (Design Leadership Board 2012).

The rise of service design in the context of urban development is primarily based on its focus on the quality of interactions and experiences, on its capacity to support the creation of networks involving a wide variety of actors and stakeholders and on the effectiveness of an experimental approach that uses prototypes to assess solutions before any further resources are committed to implementation.

These same characteristics are also sustaining the introduction of design thinking and user-centred design processes in revising the approach to policy-making in cities, integrating a bottom-up perspective into a traditionally top-down process. The emergence of policy innovation labs that are trying to integrate design methods is proof that governments (not only in cities) are waking up to the need to look beyond traditional policy-making and top-down service delivery (Bason 2010, 2014; Burns et al. 2006; Kimbell 2015; Deserti and Rizzo 2015). The multidisciplinary approach characterising these policy innovation labs is slowly overcoming the technocratic perspective: public bodies are realising that there is huge value in bringing key stakeholders and citizens together around issues that matter to and affect them. At the same time, we must observe that – despite the attempts to push for the introduction of a new approach – innovation calls for an overall cultural change, which will surely be slow, incremental and related to the transformation of well-established organisations and practices. In this frame, the introduction of user-centred perspectives and of design practices in the sphere of public services constitutes one of the major challenges to be faced.

4.1.4 *The City as a Participated Construct*

Born as an approach to the design of technological and organisational systems (sociotechnical systems) that emphasises the active involvement of the users of the system in the design and decision-making processes, participatory design (PD) has been applied to a variety of situations within the context of urban planning and community building, especially during the 1970s and 1980s of the last century.

After a couple of decades in which the prevalence of a strategic perspective has driven attention towards other approaches, today PD is being rediscovered as the most suitable approach to create the conditions necessary to set up innovation ecosystems where citizens and networks of stakeholders can co-produce solutions in partnership with public actors (Bjorgvinsson et al. 2010; Concilio et al. 2014). Researchers in the design field are arguing that contemporary PD should be interpreted in a wider frame and not only looking at the involvement of end users: when contexts are complex, PD is a promising approach to envision and develop solutions engaging with local communities and responding at the same time to the contradictory needs of multiple actors and stakeholders.

This notion of PD primarily refers to the conceptualisation of Ehn (2008) and of Bjorgvinsson et al. (2010), who propose a radical shift from the traditional view of PD that considers the object to be designed as a well-defined product or service, where the final users become active agents or codesigners, to a new definition that sees participation as key to the realisation of long-term partnerships for the sustainability of new collaborative services.

In this vision, both the object and the modes of design are new: from products designed as answers to specifications to services codesigned and co-produced to transform a social context (a city, a neighbourhood, a square, a street) facing unmet social challenges. From this point of view, the novelty that design introduces with respect to the tradition of participation in planning and urban studies (Sclavi 2000) is twofold: on the one end, the notion of design here introduced refers to the capability of constructing partnerships and networks that reside at the core of new services; on the other hand, at the micro-scale, PD is applied to concretely give shape to services and the quality of interaction with users rather than to govern decision-making processes on infrastructures, policies and regulations.

On the basis of this new notion, PD can be described as a complex and highly dynamic process (Deserti and Rizzo 2015) that can be applied to cities to generate public and collaborative services[2] (Baek et al. 2010).

4.1.5 The City as a Complex System

The city as a complex system is a well-established concept, both in studies rooted in the complexity theory[3] and in studies that introduce the concept of complexity without reference to a specific theoretical background. Nonetheless, the evolution of technologies, and particularly of ICT, somehow acted as a game changer: the modes in which different variables and agents interact in urban environments have been profoundly transformed by the introduction of ICT infrastructures and platforms, which multiply the possibilities of interaction among different subsystems and different levels of the same subsystem, making complexity grow.

The main characteristic of complex systems is the interaction of a relevant number of independent variables in interdependent and unpredictable ways. Interdependency with other systems is in fact a typical trait of many of the systems we deal with every day. In this respect, the city may be described as a paradigmatic case. We could more properly call it a "system of systems", meaning that it can be

[2]Collaborative services possess a set of characteristics that the Study on Collaborative Production in eGovernment (SMART 2010–0075) (European Commission 2012) has clearly described and analysed through 150 cases from across Europe: "Not purely bottom-up (. . .), not all about government data (. . .), applied across all services".

[3]For a critical review of the complexity theories of cities, see Portugali 2012.

interpreted as an overall organism, structured in subsystems that interact with wider systems whose main terminals are concentrated in the city itself. Transportation, work, education, healthcare, etc. are examples of such complex systems which interact with one another and whose terminals are typically concentrated in urban environments. At the same time, we should notice that urban environments do not represent only a setting but also places where new needs and challenges take shape and call for solutions. In other words, some questions may be understood and solved by only taking into account their "urban" dimension: cities, with their peculiar complexity, generate specific problems or attribute specific characteristics or intensity to long-standing transversal challenges.

When operating in complex systems, there's always a thin line between expected results and unintended consequences. Even if this is a typical problem of innovation in general, it may become particularly critical when systems are complex and the linear cause-effect relations are substituted by constant, non-linear changes in which a multitude of variables interact and modify each other. The design and the management of such systems are quite difficult and call for the adoption of new tools and the integration of knowledge across disciplines. As Irene Sanders (2008: 276) put it: *Thinking of cities as complex adaptive systems challenges us to review and revise our current planning, engineering, and design methodologies, which in most cases reflect a more linear, Newtonian worldview.* Even if the necessity to break knowledge silos has been clearly enounced for years, it still seems far from being realised. This is in our view the major challenge for design today and not only with reference to cities: that of being open to dialogue with other disciplines while at the same time being capable of displaying and making use of its own competencies. A challenge that should be overcome in the first place within (a new generation of) designers, before becoming a question of relations among disciplines; interdisciplinarity is in fact a question of revising disciplines themselves. In this sense, the application of design knowledge and practices to the transformation of cities is emerging as one of the most interesting and promising laboratories of multi- and interdisciplinarity, where the integration of different cultures of innovation may be realised. The evolution of some experiences, like that of the Milan "Fuorisalone" that we are about to illustrate, can be seen as clear representations of how the visions of the city that we briefly described coexist and are connected with the steady evolution of the culture of design.

4.2 Milan Fuorisalone: An Exemplary Story of the Relationship Between Design and the City

4.2.1 A Brief History

Milan "Fuorisalone" (literally, outside the exhibition), now renamed "Milan Design Week", is one of the most important international design events, which involves the entire city of Milan in a frenzy of events and installations. Together with the annual furniture exhibition, it attracts people from all over the world and has become the major business event taking place in Milan.

Born in the 1970s along with a few side initiatives organised during the annual furniture exhibition and held in the showrooms of very few furniture brands, Fuorisalone started growing in the 1980s and had a push in the 1990s.

Initially, some furniture brands started to independently organise presentations of their new collections in their in-city showrooms. In 1991, Comitato Organizzatore del Salone del Mobile Italiano (COSMIT), the organiser of the official exhibition, moved the 30th edition from September to April. To fill the space left open in the calendar in September, the *Interni* magazine launched the first design week, in the form of a set of events for the presentation of new products taking place in a network of in-city furniture showrooms, and published a leaflet that would have become the first guide to Fuorisalone (Cuman 2012). This attempt of creating an alternative and fully autonomous business came to an end after the 2nd edition, when the design week was realigned with the new calendar of the official exhibition, becoming de facto its side event and assuming the name "Fuorisalone".

4.2.2 A Bottom-Up Multifaceted Initiative

Compared to the official furniture exhibition, Fuorisalone was thus born as a bottom-up initiative, characterised by a loose ownership, which drove most of the institutional actors to look upon it as a non-legitimate competitor. Fuorisalone actually appeared from the very beginning as a multifaceted initiative, with a mix of cultural and commercial activities: on the one hand, it hosted the more experimental work of young designers and companies who could not find place in the official exhibition; on the other hand, it took the form of a parallel commercial exhibition in which some of the established companies found it more convenient to use their showrooms or other in-city venues rather than renting spaces in the official exhibition buildings.

Fuorisalone appeared more as a social than a business event: showrooms used to be (and still are) open after hours, offering an "aperitivo" and organising parties, addressing the whole community gathering around the creative professions – and to a growing number of common citizen and visitors – an opportunity to meet, exchange ideas, network and have fun. At the same time, despite its "social" nature, the growth of Fuorisalone as a business was constant. While estimating it is quite difficult, as there are many operators and the borders between the core and satellite activities are quite blurred, the growth of the number of events can give an account of its expansion: where in 1991 there were 50 events with one organiser, in 2015 there were nearly 1,250 events with a multiplicity of organisers.

Today Fuorisalone involves a multitude of subjects, operating in a regime of "co-opetition" (Brandenburger and Nalebuff 1996). This multifaceted nature makes it quite different from the many design weeks that have been established and are being established at a quite fast pace around the world. While most of these events are "compact" – meaning that they assume the nature of a fair with an official organiser (or at least a structured group of organisers) and a target market, as a model – Fuorisalone is to some extent the opposite of a fair: there is no official organiser (nor many official organisers), no brand (nor many sub-brands) and even no official name (nor many official names for the different sub-brands).

4.2.3 The Relation with the City

The engagement of the city is a key characteristic of Fuorisalone. Milan is not only the setting of the event but also its motor. Fuorisalone depends on the city, but at the same time, it acts on it. In this sense, Fuorisalone is at the same time a representation and an occasion of transformation of the city: an agent that provokes transitory and permanent changes. The areas where it takes place are temporarily transformed by the events, which have a visible impact on the tangible characteristics of the interior and exterior spaces, the number and kind of people visiting the spaces and the intangible atmosphere resulting from the interaction between people and spaces.

The different areas, with their identities and vocations, work as a stage, but the events and the installations turn them into something different. The locations assume a peculiar atmosphere that makes them special and attractive, allowing many hidden places ready to be discovered not only by the growing number of foreign visitors but also by the citizens themselves.

Discovery is actually a fundamental element of Fuorisalone's value proposition, as the unveiling of new products and design trends goes along with the unveiling of spaces. The scattered geography of the events, the difficulty in finding places, the crawling of crowds between different districts, the awareness that seeing everything is simply not possible, the emotion of just stumbling into something unexpected and the pleasure of getting somewhere due to word of mouth are all elements of the Fuorisalone experience.

An interesting aspect is the relation between the transitory nature of the event and the long-term change of the city. The transformation that Fuorisalone brings is not only temporary: its presence has contributed to building an overall imagery on top of some of the areas where it takes place (and actually on top of the whole city), which becomes the actor of their long-term change and appropriation by the expanding creative communities.

This phenomenon is not original per se, as it has taken place in other cities around the world in connection with operations of urban renewal and branding (Anholt 2007): where creative communities have sometimes been used as largely unaware actors in valorisation processes of real estate, often leading to gentrification and the expulsion of the vulnerable social classes (Cameron and Coaffee 2005). Even if we may recognise some of these negative effects in the case of Milan, what happened with Fuorisalone and the creation of the Milanese design districts is quite far from other much more "artificial" processes of transformation of contemporary cities. In the case of Milan, the construction of the design districts was based on a real vocation of the city, historically acting as a strategic knowledge hub for a wide system of manufacturing clusters located in the Lombardy region and in the whole country (Maffei and Simonelli 2002). The city's districts build on this overall character and on their specificities: some are more bourgeois, residential and commercial; others have a more ex-industrial and now tertiary identity. Not only do the operators managing events in these districts act as competitors, but the same could be said about the districts themselves as collective entities: each district fights its way to become the "place to be" in the next edition of Fuorisalone. This perspective is in line with some already described characteristics of urban development, where single areas within the same city may be seen as competitors (Mäding 2006), or as actors that operate in the already-cited regime of co-opetition.

4.2.4 The Transformation of the City

Fuorisalone is actually configured as a multicentre event, taking place in different districts and locations, whose geography is in constant evolution. New entrants challenge the established organisers, districts and locations: each of them tries to develop its own value proposition and identity, based on the vocation of the territory and on the specificities and capacities of the organisers.

The background of the organisers largely hails back to design culture and integrates skills that are in constant evolution. Their modes of action provide a lively representation of how the different visions of the relationship between design and the transformation of the city coexist and interact with each other.

Initially, the main players were specialised design magazines, such as *Interni*. Their business model (Perkmann and Spicer 2010; Teece 2010) was based on the synergies made with their publishing activities: they edited, printed and freely distributed to visitors a pocket guide of the events, through which they could sell advertising spaces to the exhibitors, from a simple citation to a more visible presence in the guide. At the end of the 1980s, when a few new entrants started publishing alternative guides, *Interni* put together its publishing activity with the organisation of events or rather expressions of a direct cultural presence in the form of installations in public or semipublic spaces that after a few years found their stable venue at the cloisters of the University of Milan. To these actors, Fuorisalone represents a form of diversification of the core business and a way of creating value by leveraging the event and its international fame. The dynamics of business diversification through brand management and their connection to the exploitation of the overall brand of the city through "unofficial" forms of co-branding can be easily retrieved in the characteristics of the business model of these operators. In the first step of the evolution of their business model, they tried to create sub-brand architectures or "branducts", exploiting their guides and registering umbrella brands for the whole Fuorisalone (such as *Interni*'s "Milan Design Capital"), interpreting the city as a brand to be developed and exploited. Today they exploit Fuorisalone as a lever for their core business and – due to the crisis of the traditional publishing industry – as a means to enter the field of event planning and management, as a promising side business where they can exploit their network of relations and integrate their core knowledge on publishing and advertising. In this direction, we can interpret also the entrance of foreign players such as Wallpaper.

Other organisers come from the communication design field: Fuorisalone is at the core of their business, and they exploit it primarily by creating area brands on top of which they can build an integrated offering of services. In this case, the design and the offering of services becomes the core business, but place branding is still a relevant initial asset. The brand of the organiser tends to be invisible, or to appear with a lower level of visibility, while all the branding strategies and operations are focused on branding the areas of the city that they "control". We can take Studiolabo – the business entity behind fuorisalone.it and Brera Design District (run by two graduates in design at Politecnico di Milano) – as the benchmark of these kinds of companies. On the one hand, Studiolabo operates as a media company. In this branch of the business, revenues come from the sales of advertising spaces in multichannel media platforms, in a variety of formats for all the digital media and the printed guides to the events: these spaces may be part of the offering to the sponsors, but may also be sold as separate packages, as done in media broadcasting and in traditional publishing. On the other hand, Studiolabo operates as a service design company and as an intermediary in a multisided market (Caillaud and Jullien 2003; Rysman 2009). The difference compared with the traditional business model of a fair, where the venue is conceived from the very beginning to host the business, is that here the transactions require a much more complex operative structure. In the majority of the cases, locations do not belong to the organisers, nor do organisers control just one big venue that can be subdivided and allotted. In many cases locations are spaces normally hosting other activities that can be temporarily used to show products and installations. The variety of these spaces is impressive: local shops, factories and warehouse converted into tertiary spaces, small laboratories, private houses, bars and restaurants, hotel lobbies, institutional venues, public squares, etc. While for some of these spaces there might be a direct relation between the owner and the exhibitors, in the majority of the cases, the organisers provide an intermediation service, whose importance has grown in correspondence with the progressive internationalisation of the exhibitors. This service, even if theoretically similar to that of a fair, requires much stronger networking skills and the capacity of showing the potentialities of the locations to faraway prospect clients. Due to these peculiar characteristics, it has taken the shape of a side business, run through digital platforms (a typical character of two-sided business models), such as Milano Location and Brera Real Estate for Studiolabo and Tortona Locations for Tortona Design Week. In all these cases, the business involves and exploits local resources and is based on the capacity of creating networks and partnerships to co-produce value.

The co-production logic is often at the core of the modes of value creation: for example, fuorisalone.it was conceived from the very beginning (before the era of social media) as a crowdsourcing platform, where independent "design enthusiasts" – primarily young designers and students of design schools – were engaged as reporters to cover the many scattered events, feeding the platform with live updates and reports of what was happening.

One of the relevant issues that these organisers face is that of "unseasoning" their business, making the event live throughout all the year. Pursuing this goal, operators like Studiolabo concretely act on the city in order to tangibly and intangibly transform the urban areas where their events are located into permanent locations for the design activities, enforcing and speeding up the spontaneous long-term transformation of those areas into design districts or into privileged locations for the cultural and creative industries. This strategy is quite well expressed in the Brera Design District's tagline "The best of design all year round". Concretely, this means on the one hand finding permanent locations and on the other hand organising activities beyond Fuorisalone.

4.2.5 The Spin-Off Activities

Another interesting phenomenon is that of the spin-off activities and businesses. Fuorisalone has a relevant direct impact on the economy of the city, but it is also the driver of specific initiatives that build on the presence of an enormous number of visitors and on the mobilisation of people during the design week. Among others, we can describe Elita as a paradigmatic case. Born as a non-profit association organising a music festival during Fuorisalone, Elita has subsequently created a for-profit entertainment company, organising the Design Week Festival and launching side initiatives, such as the ExtraSmall designer's market and – again in the perspective of "unseasoning" the business and of leveraging on the community – the Elita bar, a permanent place for the gathering of the creative community. It is interesting to note that the Elita Design Week Festival, although different in what is shown, assumed the same networking and pervasive character of the other Fuorisalone events: initially (and still) headquartered at Teatro Franco Parenti, it is expanding throughout the city involving a growing network of music clubs.

4.2.6 The Relation with the Institutional Frame

Another interesting aspect is the interconnection of single operators and of their specific business models with the institutional frame.

From a historical analysis, it is clear that Fuorisalone was largely built outside institutional boundaries. In this, the difference with most of the design weeks popping up all over the world is huge. Apart for some sponsorship and other minor efforts, local institutions did not play an effective role for years and simply observed the surging phenomenon. At a certain point, it became so huge that it was simply impossible to not take stock of it, but still the competition between the official exhibition and the independent Fuorisalone was a barrier to an explicit institutional support. Today – since all the players realised that they are not only competing among themselves but also in the much larger international arena – institutions are actually trying to support Fuorisalone as a fundamental element of the overall value proposition of the annual furniture exhibition and as one of the elements of qualification of the city of Milan as the major international design hub. The Milan municipality, as well as other public institutions, is thus trying to smooth the bureaucratic processes behind its realisation in order to foster cooperation between the multiplicity of actors involved and to combine the official fair and the "unofficial" Fuorisalone into a single value proposition. In all this, the umbrella brand "Milan Design Week" is still much weaker than the single sub-brands that it is supposed to cover, to the point that the umbrella brand itself has a name but not a defined visual identity nor a subject really taking control of it. This is leading to a quite complex process of negotiation and alignment: due to the ways in which Fuorisalone took its current shape, institutions seem to be aware that the alignment of the whole system cannot be taken in the perspective of its management, but in that of its governance.

4.2.7 The Twofold Role of Design

In all this, design plays a twofold role. On the one hand, design is the content of the exhibition: it is what is shown and what people come for. On the other hand, design is the intangible culture behind the event (Julier 2013): it is the knowledge that gives shape to the event as it is. In this sense design is intended in its larger meaning: a culture expressing or underpinning a special way of doing things. Design is a fundamental character of the culture of Milan, which may be retrieved at all the levels in which the culture of organisations has been articulated: visible artefacts, explicitly espoused values and invisible underlying assumptions (Schein 1999). Design permeates the city, with its visible pervasive presence, but also with its invisible processes, values and beliefs. The evolution of Fuorisalone, and the transformation of the city that it brought both at the tangible and at the intangible levels, can be thus described as led by design thinking (Brown 2009; Lockwood 2009) and culture (Deserti and Rizzo 2014; Concilio et al. 2014) as pervasive characters of the city.

Design is also the engine of innovation of the event. Its unique competitive positioning in the international markets builds on design-driven innovation (Verganti 2010), or else on the combination of technological innovation (in particular an advanced use of digital technologies and platforms) with an innovation of meaning, based on the specificity of the cultural environment in which the event takes place and on its relation with the material substrate of the city.

4.2.8 Conclusions

The modes of operation of Fuorisalone may be described as similar to those of an industrial cluster. The phenomena of cooperation, competition, innovation, imitation, knowledge creation and exchange are actually the same that have been described in industrial clusters (Porter 1990). The transformation dynamics of Fuorisalone in its steady internationalisation process are also similar to those observed in industrial clusters during the globalisation of competition: networks remain anchored to their local core but become much larger in their geographical base, while competition becomes international, calling for the capacity of connecting a situated know-how with international markets and the local culture with the global trends. Fuorisalone, and actually the whole design network underpinning it, works as a tertiary district characterised by an intense interaction of actors: operators cooperate, compete, innovate and imitate each other, advancing overall knowledge and generating value.

The mix of culture and business resides at the base of its value proposition, but at the same time, it poses a dilemma to the organisers, challenging the business's sustainability. Visitors are in search of a different experience than the one that they can have at the commercial fair: experimental designs, perspectives on new trends, emotional installations and relaxed networking. Organisers must provide all this, but they have to find ways of making it economically viable: they manage the constant tension and trade-off between being commercial and incrementing revenues and being experimental and cutting-edge. This tension may be interpreted as mostly similar to that occurring in organisations striving to combine exploration and exploitation (Martin 2009) or to manage product portfolios where sheer economic performances must be combined with other ratios. The organisers of Fuorisalone – typical design-driven companies – can be described as ambidextrous organisations (March 1991), as they must combine the capacity of constantly innovating their offering with that of honing processes to create conditions of efficiency for economic exploitation. In this sense, Fuorisalone itself paradigmatically represents some of the unvarying dilemmas posed to design, while its evolution describes the constantly changing ways in which they can be faced, and reflects the profound transformation of the ways in which design is applied to cities or, vice versa, of the ways in which design can contribute to the transformation of cities.

References

Anholt S (2007) Competitive identity: the new brand management for nations, cities and regions. Palgrave Macmillan, London

Baek J, Manzini E, Rizzo F (2010) Sustainable collaborative services on the digital platform. Proceedings of Design Research Society Conference. DRS, Montreal, 9–12 July

Bason C (2010) Leading public sector innovation: co-creating for a better society. Policy Press, Bristol
Bason C (2014) Design for policy. Gower, Farnham
Bjorgvinsson E, Ehn P, Hillgren P-A (2010) Participatory design and democratizing in-novation. In: Proceedings of the 11th biennial participatory design conference, ACM, New York
Brandenburger AM, Nalebuff BJ (1996) Co-opetition. Doubleday, New York
Brown T (2009) Change by design. How design thinking transforms organizations and inspires innovation. Harper Collins, New York
Burns C, Cottam H, Vanstone C, Winhall J (2006) Transformation design. RED Paper 0.2, Design Council, London
Caillaud B, Jullien B (2003) Chicken & egg: competition among intermediation service providers. RAND J Econ 34(2):309–328
Cameron S, Coaffee J (2005) Art, gentrification and regeneration – from artist as pioneer to public arts. Eur J Hous Policy 5(1):39–58
Concilio G, Deserti A, Rizzo F (2014) Exploring the interplay between urban governance and smart services codesign. ID&A Interact Des Archit 20:33–47, ISSN: 1826–9745
Cuman A (2012) MediaSpaces, urban events and mobile experience: an ethnographic enquiry into the social production of the city of design, Doctoral dissertation, Università Cattolica di Milano, Milano
Deserti A, Rizzo F (2014) Design and the cultures of enterprises. Des Issues 30(1):36–56
Deserti A, Rizzo F (2015) Design and organisational change in the public sector. Des Manag J 9(1):85–97
Design Leadership Board (2012) Design for growth and prosperity. European Commission, Brussels
Directorate General for Internal Policies, European Parliament (2014) Mapping smart cities in the EU. European Commission, Brussels
Ehn P (2008) Participation in design things, proceedings of the 10th biennial participatory design conference. ACM, New York
European Commission (2012) Study on collaborative production in eGovernment (smart 2010–0075). Part 1. Practice and implications. European Commission, Brussels
Gold RJ, Ward SV (1994) Place promotion: the use of publicity and marketing to sell towns and regions. Wiley, Chichester
Julier G (2013) The culture of design, 3rd edn. Sage, London
Kimbell L (2015) Applying design approaches to policy making: discovering policy lab. University of Brighton, Brighton
Kotler K, Haider D, Rein I (1993) Marketing places. Free Press, New York
Landry C (1990) Glasgow: the creative city and its cultural economy. Glasgow Development Agency, Glasgow
Landry C (2000) The creative city: a toolkit for urban innovators. Earthscan, London
Landry C, Bianchini F (1995) The creative city. Demos, London
Lockwood T (2009) Design thinking: integrating innovation, customer experience, and brand value. Allworth Press, New York
Mäding H (2006) Cities and regions in competition – an outline. Deutsches Institut für Urbanistik, Berlin, http://www.difu.de/publikationen/cities-and-regions-in-competition-an-outline.html. Accessed 20 Oct 2015
Maffei S, Simonelli G (2002) I territori del design. IlSole24Ore, Milano
March JG (1991) Exploration and exploitation in organizational learning. Organ Sci 2(1):71–87
Martin R (2009) The design of business. Why design thinking is the next competitive advantage. Harvard Business School Publishing, Boston
Morgan N, Pritchard A, Pride R (eds) (2002) Destination branding: creating the unique destination proposition. Butterworth Heinemann, Oxford
Perkmann M, Spicer A (2010) What are business models? Developing a theory of performative representations. Res Sociol Organ 29:265–275
Porter ME (1990) The competitive advantage of nations. Free Press, New York

Portugali J (2012) Complexity theories of cities: achievements, criticism and potentials. In: Portugali J, Meyer H, Stolk E, Tan E (eds) Complexity theories of cities have come of age. Springer, Heidelberg, pp 47–62
Rysman M (2009) The economics of two-sided markets. J Econ Perspect 23(3):125–143
Sanders IT (2008) Complex systems thinking and new urbanism. In: Haas T (ed) New urbanism and beyond: designing cities for the future. Rizzoli, New York, pp 275–279
Schein EH (1999) The corporate culture survival guide. Jossey-Bass, San Francisco
Sclavi M (2000) Arte di ascoltare e mondi possibili: come si esce dalle cornici di cui siamo parte. Le Vespe, Pescara
Teece DJ (2010) Business models, business strategy and innovation. Long Range Plan 43(2–3):172–194
Vargo SL, Lusch RF (2004) Evolving to a new dominant logic for marketing. J Mark 68(1):1–17
Verganti R (2010) Design driven innovation. Harvard Business Press, Boston

Chapter 5
Design At the Intersection Among City Challenges, New Public Services, and Policy-Making

Francesca Rizzo

Abstract Societies and cities are living in times of deep cultural changes. Design of course has sought to tackle city societal problems in the past, but this has largely been confined to design activism (and to the political sphere of the design action). Its main aim has been to raise awareness of specific problems and/or demonstrate dissent with mainstream cultures. In the last 10 years, however, there have been various design initiatives that have worked at city scale to solve societal challenges, producing reliable and useful solutions and valuable impacts on the life of real people.

When applied to city challenges, design assumes the practice of complex participatory processes involving a large number of actors and stakeholders in tense settings or open conflicts.

Complex participatory processes go beyond the established principle of designing for context-dependent problems, extending the idea of participation to include (1) the relation between the context of the problem to be addressed and the design of the network that will coproduce the solution and (2) testing different configurations of that network until a robust partnership is individualized and established in some institutional form.

The contribution relies on the intuition that design can act as an agent of change for public institutions, which are currently facing new and unmet societal challenges that appear to affect cities at different levels. These include the quality of the services offered by municipalities and the way in which public institutions deal with service innovation in conditions of scarce resources, with new phenomena such as social innovation. The chapter also introduces a design-led project implemented in the framework of the My Neighborhood European Project, with a double aim, i.e., to experiment service design as a tool for designing innovation in the public sector and to experiment service design as a tool to boost innovation in the culture of a public institution (the Municipality of Milan).

F. Rizzo (✉)
Department of Architecture and Territorial Development, University of Bologna,
Via Risorgimento 2, 40137 Bologna, Italy
e-mail: f.rizzo@unibo.it

G. Concilio, F. Rizzo (eds.), *Human Smart Cities*, Urban and Landscape Perspectives 19,
DOI 10.1007/978-3-319-33024-2_5

Keywords Complex participatory design • Service design • Collaborative service • Social innovation

5.1 Design for City Challenges

This is not just an age of recession and austerity, but of truly profound disruption and dislocation, so the kinds of challenges that cities face at this time are only seen properly in that context (Hermant-de-Callataÿ and Svanfeldt 2011). It is an economic disruption, because cities in Europe, for instance, are desperately searching for sustainable economic models, and, even when growth returns, the question of which economic model best sustains society will still not be settled.

It is a time of profound technological disruption, with a continuous wave of new technologies and applications, which are destroying old industries and providing new opportunities to create value and new ways of collaborating at the same time.

Governments everywhere, at every level, from cities to regions, from nations to the whole European continent, are facing serious challenges because of their distance from society and their ineffectiveness and inefficiency at addressing important issues such as the need for growth (these problems threaten to overrun existing infrastructure and create profound inequalities). In industrialized contexts, governments are not dealing with the problem of decline and decay, which are associated with economic disinvestment that leads to social and political disinvestment. To crown it all, societies and cities are living in times of deep cultural changes, brought on partly by technology, which is a growing culture of massive participation in which people, citizens, and creative communities increasingly expect to be part of, to voice their opinions, to connect, to organize, to choose, and to have feedback on absolutely everything they are involved with.

Considering all the above, it is possible to see why the challenge for the next 20 years (at least) is not to develop new products or new services – it is to develop new urban ecosystems with new cultures and new forms of behavior. These urban ecosystems have to be both more efficient and more human oriented, to guide planners, policy-makers, service designers, and intermediaries through the challenge of making cities capable of rising to their internal societal challenges, using, among other things, the advantages provided by technology and the way they can change our lives (Mulgan and Leadbeater 2013).

This is the example of the city of Curitiba in Brazil, one of the most impressive smart cities in the world. Curitiba is a city which is designed in a systemic way; it has plans for transport, for building, and for regulations, but it also understands that its future depends on mobilizing its citizens to create solutions. In a city like Curitiba, the smart transport system has been combined with civic spaces that are open, easy to access, and convivial. Here the solution to some problems like recycling depends not on large systems but on creating a lot of micro recycling entrepreneurs, who create a business out of collecting rubbish and then recycling it. Models of cities like Curitiba are questioning the existing planning methods used by governments to provide for citizens when dealing with city challenges. How do we deal with a growing environmental crisis that fundamentally questions the sustainability of our way of life (making us more aware of its effects on social well-being) and raises the question of pollution and of energy resources? How do we seriously address the demographic situation and the challenge that it presents, i.e., dealing with a much higher number of elderly people and limited resources to take care of them? All these questions are co-defining the current state of crisis, and they seem to imply that, through a "failure of agencies," public institutions and organizations have failed to answer them.

Design of course has sought to tackle city societal problems in the past, but this has largely been confined to design activism (and to the political sphere of the design action). Its main aim has been to raise awareness of specific problems and/or demonstrate dissent with mainstream cultures. In the last 10 years, however, there have been various design initiatives that have worked at city scale to solve societal challenges, producing reliable and useful solutions and valuable impacts on the life of real people. Examples include John Thackara's Doors of Perception conferences and blog (http://www.doorsofperception.com), Bruce Mau's book *Massive Change* (2004) and his Institute Without Boundaries, Cameron Sinclair's nonprofit organization Architecture for Humanity (http://architectureforhumanity.org), and Emily Pilloton's Project H Design: Supporting Community (http://ced.berkeley.edu/frameworks/2012/project-h-design/). What is emerging today is a new wave of design projects, a new vision – partially elaborated in the context of design culture – that focuses on the application of design to tackle city challenges as complex problems and to find solutions at the level of the cities and territories (Latour 2010; Binder et al. 2011; Hamdi 2004).

Design culture has received many definitions in literature (Julier 2006). The notion here embraced is the one that pushes on the front of the capability of design to work from a context-dependent point of view, together with the design attitude to scale solutions through experimentation (Deserti and Rizzo 2014).

When applied to city challenges, this notion assumes the practice of complex participatory processes involving a large number of actors and stakeholders in tense settings or open conflicts. Complex participatory processes go beyond the established principle of designing for context-dependent problems, extending the idea of participation to include (1) the relation between the context of the problem to be addressed and the design of the network that will coproduce the solution and (2) testing different configurations of that network until a robust partnership is individualized and established in some institutional form.

Many projects in particular are evidence of this new kind of design practices (Manzini and Rizzo 2012; Julier 2013). They exemplify a new trend in the relation between design and cities – design is definitively moving toward the design of complex systems, gaining a position as a discipline that can have an impact on society and on the real life quality of people, as a culture of innovation that transforms cities and territories.

Malmo Living Lab is an example of this kind of design-led project, which works to boost micro and bottom-up solutions as services to address local city challenges (Björgvinsson et al. 2012). In 2009 the media institute Medea at Malmö University (with the financial support from the KK Foundation and EU structural funds) launched three living labs for coproduction and social innovation in the city of Malmö. The city is characterized by multiethnicity, cultural production, youth culture, and new media industry. This is also the rationale behind the content orientation and the cultural and geographic position of the three suggested living labs. "The Neighborhood" is set up in the contentious multiethnic Rosengård suburb and focuses on changes in urban space, collaborative services, and social media. "The Stage," set in the vibrant club, music, theater, and subculture environment around Möllevångstorget, focuses on cultural production and cross media. "The Factory" is located by the Stapelbäddsparken skateboard arena in Västra Hamnen, in the heart of the city's new media cluster, focusing on innovation strategies where users can develop fully functional prototypes in an open-source and mixed-media environment. Each living lab is carrying out a series of self-standing projects on context-dependent challenges, with the help of new technologies. Projects act in order to activate a series of services, inspired by social innovation solutions, to regenerate the urban communities that live in these places. The final aim of the project is to help people design micro solutions for their problems.

Designs of the Time (Dott) is an example that is the opposite of the Malmo case study, because it represents a top-down strategy for introducing design as culture of innovation for cities and regions. Dott07 and Dott09 are the two editions led by the UK Design Council, a highly qualified actor in the field of design, operating from 2006 to deliver solutions for societal challenges embedded in UK urban areas. Dott07, which started in 2007, was delivered by the regional development agency for the North East of England (One North East), without the involvement of the central government. The first year of the program consisted in evaluating current community initiatives in the region, and seven core projects were short-listed for in-depth action from a list of 200 projects. In the second year of the program, the design teams examined new tools and platforms for creating sustainable and innovative solutions to complex societal problems through design. The seven short-listed projects were Alzheimer100 (dementia), DaSH (sexual health), OurNewSchool (building new schools), Low Carb Lane (domestic energy), New Work (for improving the day-to-day experiences of SMEs), Move Me (rural mobility), and Urban Farming (exploring local food systems). The first Dott was led by Program Director John Thackara. The second and final program, carried out in Cornwall during 2009–2010, was led by Andrea Siodmok. The key attributes of both programs were a series of 20 citizen projects, a skills program, and policy recommendations. The Design Council, Cornwall Council, University College Falmouth, and the Technology Strategy Board were partners behind Dott in Cornwall and the Isles of Scilly during 2009–2010. Dott worked with citizens to help them co-design solutions to the daily problems they faced. The challenges explored included what design can do to reduce ecological footprint, how to improve directions to new employment opportunities relevant to the economy and society of twenty-first-century Cornwall, how to help older people find new positions and stay in their jobs for longer, and how to use design to help push quality Cornish produce into the local and global spotlight. Although most of the individual projects were small-scale, they were real-life examples of communities teaming up with designers to create practical solutions to important issues affecting their living environment.

This project proposed a design-driven amplification method in order to improve and expand the capacity of territories and communities to recognize and solve local problems, using design to envision solutions. Dott can be described as a series of creative and grassroots community design projects, intended to improve design awareness at different levels of society and to stimulate social, environmental, and economic innovation in a local territory (Tackara 2007).

Projects like these are challenging the traditional design project format and are based on building long-term working relations between diverse groups and actors in society and on encouraging mutual learning between them (Di Salvo 2012; Brown 2009; Brown and Wyatt 2010). At a practical level, they emphasize the processes that typically start by connecting to diverse grassroots organizations in a territorial unit (a neighborhood, a rural area, a street, a square) and creating an understanding of their ongoing everyday activities and of how these could potentially support service and social innovation. In terms of vision, they stress the importance of allowing a plurality of voices, a flexible allocation of resources and time, how experimentation and innovation can emerge from the continuous matchmaking of diverse actors, and their needs and competences (innovation as a set of continuous and ongoing relational processes, rather than as a fixed technical system).

Although the notion of starting from the context is often seen as starting from the physical spaces people live or operate in, in this new wave of projects, it gains a new meaning, especially in connection with social innovation (Murray et al. 2010; Norman 2012). In addition to looking at the material aspects of innovation (first step), these projects also try to create an understanding of existing and potential social relations in the area, by mapping the actors and the existing initiatives that can become potential resources for innovation. The second step is to consider connections to other initiatives in the city by identifying common issues and joint experiments. Social capital and learning between disparate groups will increase through these experiments and new social infrastructure is built (Latour 2005, 2010). Projects like these are pushing design to the forefront of innovation of public services and policy-making.

5.2 Design for Public Services and Policy-Making

Although the current types of city challenges and problems represent new opportunities for design, they also illustrate the limits of welfare services bound in twentieth-century models, based on the assumption that the state delivers services to passive citizens or commissions specified solutions to well-defined problems. Some of the most urgent and costly challenges facing welfare systems are those that require an understanding of the personal, contextual, and invariably multidimensional aspects of people's real lives. Others require types of services that are able to engage and collaborate more productively with people, while others build on individual and social assets to create useful change.

Design is now commonly seen to have important contributions to make in helping public organizations face these challenges, which is testified by the number of "labs" that have been set up across the world. The purpose of these labs is to introduce an experimental approach to building knowledge and creating system change, to address the challenges facing governments and citizens. This is pushing design into the upper echelons of governments, right into the systems, institutions, and rhetorics of public organization across the world.

Different projects and programs are paying more attention to systemic levels and are trying to explore how design could potentially have an impact on larger systems and, especially, how design could reach into the public sector and into municipal offices (Bason 2010; Christiansen and Bunt 2012; Botero and Saad-Sulonen 2013; Manzini and Staszowski: 2013; Deserti and Rizzo 2015).

In 2012, the European Design Leadership Board's Design for Growth and Prosperity report called for designers in residence in EU institutions and member states to increase the design skills of public sector administrators, so that design methods could be used for effective policy-making. In the meantime, in the United Kingdom, the *Civil Service Reform Plan* (HM Government 2012) recognized the value of lab working methods, quoting MindLab in Denmark as an example of the approach. The APDIG Design Commission's *Restarting Britain: Design and Public Services* report (UK Design Commission 2013) recommended that the Cabinet Office should take responsibility for developing design skills across the government, specifically trailing a multidisciplinary design studio method for policy originating and for a wider drive to equip policy-makers with design skills.

In the United Kingdom, several service design companies, including Livework, Engine, and ThinkPublic, began to apply design methods to social and societal challenges in the early 2000s.

Furthermore, other public institutions in the United Kingdom have started projects involving design companies. For example, the Engine design firm has supported the Kent County Council in designing a new platform for co-creation, whereas the Livework design agency has created services to support hard-to-reach unemployed people. ThinkPublic, a service design company, has used participatory design to engage citizens in identifying challenges and co-designing responses to improve community health and well-being.

This occurred, among other things, with support from the Design Council (2008, 2012), which encouraged the exploration of design use in new social and public contexts through research and demonstration projects. One example was the Design Council's RED research unit, which consisted of professional designers and professionals from disciplines such as policy analysis and social sciences. Burns et al. (2006) described the RED unit's approach, called *transformation design*, as based on involving heterogeneous stakeholders from the beginning through participatory design.

In April 2014 the UK's first Policy Lab was launched. The Lab's theory of change is that using design principles to approach complex problems can result in better outcomes and that training policy-makers in design research methods, including using or commissioning ethnographic research, has the potential to transform the way that policy is made in the government.

There is growing interest in design for public services in the United States as well. It is among the areas of expertise offered by the design firms IDEO, Continuum, and Frog Design. In particular the DESIS Lab in the New School for Design in New York is working in a more European way, by helping local creative communities to develop collaborative services and sustainable lifestyles. In the research program known as Public & Collaborative NYC, the DESIS Lab explores what role design can play in building bridges between city government and people in the creation of social innovation (Staszowski et al. 2013).

An example of this new role for design is the case of "La 27e Région" developed in France. It represents one of the most interesting design-led projects developed by an independent organization that explores, in partnership with the central government (which is funding the initiative), how design approaches can influence policy development in the public sector in France. In 2008, La 27e Région started its activities as a not-for-profit "do-tank," working in, with, and for the 26 French regional governments.

La 27e Région's strategy, which has been characterized as friendly hacking, builds on embedding multidisciplinary teams, including designers, who can empower civil servants in diverse public organizations for shorter or longer periods. They use the term "hacking" because it "signifies the intent to challenge the robustness of public policy instruments" (Jégou et al. 2013, pg. 6). The final aim of the project is to disseminate design knowledge competences and skills to empower people.

They have all been working closely with a growing community of service designers, architects, and sociologists gradually engaged in design for public policy. After a first set of 16 tests spread all across the country and dealing with issues such as education, health, mobility, democracy, and procurement ("Territoires en Résidences" program, started in 2009) and after four prototypes of "public innovation labs" embedded in regional governments ("La Transfo" program, launched in 2011), La 27e Région is now entering a new phase. In partnership with the French state and local authorities who have decided to unify their efforts, La 27e Région has set up a new 4-year program, Re-acteur Public, led by a large consortium of different public institutions. Re-acteur Public is designed as a vehicle to scale up the methods, processes, and thinking developed during the past 6 years. Re-acteur Public was officially launched in May 2014 by the French Minister of Reform, together with representatives of French public administrations (both local and national). The consortium aims to explore four areas of work until 2017: skills (building design capacities among civil servants through training sessions), community (bringing together all the practitioners of design for public policy), future (exploring possible futures of the administration), and publishing (launching a collection of books, cases, and prototypes) to scale up the methods, processes, and thinking developed during the past 6 years.

During 2014 La 27e Région has also established a collaboration with the National School of Administration (Ecole Nationale d'Administration), through the Reacteur Public project, so that they can include design for policy in their programs.

However, it is clear that all these initiatives and programs start from working on the implementation of design in the public sector, by designing and redesigning public services with the aim of improving the life quality of people and suggesting a social role for design. Moreover, they represent attempts to apply design on policy as a dynamic object.

In these projects, design is playing a more transformative role, which argues for challenging established structures and triggers changes in public organizations and in how they produce policy, instead of focusing on productivity, efficiency, users' experience, or improving services within existing societal structures (Deserti and Rizzo 2015; Botero and Saad-Sulonen 2013; DiSalvo 2012).

Projects like these show how fruitful design is with regard to planning policy, since it takes the dynamic relationship as the premise in their development (Christiansen and Bunt 2012; Brown and Wyatt 2010; Burns et al. 2006). Policies can no longer be seen in their own right, but only make sense when seen in relation to their practical outlook and consequences. Unlike the traditional understanding of policy-making and governance as the rational development of models, design is predisposed to more iterative creation and stewardship, closing the gap between model development and implementation. Design as a discipline is also more at ease with complexity and uncertainty and is therefore commonly used as an innovation method. Though oversimplified, one of the core strengths of a design approach is that it starts from understanding the architecture of the problem, focusing on the actual causes and consequences involved, as well as on the interconnected systems and networks involved in dealing with it. Taking on different perspectives, asking new questions and reframing challenges can introduce innovation into thought or action processes by creating a tension with common interpretation. By asking different questions, a design approach can point to different trajectories for addressing the problem.

Moreover, this new wave of projects is disseminating a new view on policy-making as "experiments in progress." The application of design culture to policy-making remains an untapped opportunity in relation to policy and decision-making. However, the current trend for involving designers and design-based approaches in public service planning to deliver public policies creates new opportunities, which in itself is a huge opportunity to embed design into the policy-making process. A lot of work has to be done to find ways to measure the induced changes and their impact effect, at service level and at organization level.

Thus, the idea of experimentation in relation to public governance and policy development has risk connotations. To a large extent, this is understandable, given the important responsibility of ensuring public accountability and civil rights through trustworthy bureaucratic procedures and structures. Therefore, innovation, which has an unknown and unpredictable outcome, is seen as risky in contrast to known, predictable outcomes (and familiar failures) of current practices, whether they are successful or not. As a consequence, much innovation still tends to be carried outside of the core operations of public organizations. Design culture and methods help to create a legitimate space for experimentation that contains risks and expectation and supports learning from (low-cost) failure where the cause of a problem is unknown or where practices are still evolving. This is different from running an initial pilot prior to launching a full program, which is often how public policies are developed (and which has its own risks). When pilot projects have a high profile, political capital, and considerable investment, failure can come at a considerable cost. The expectation from experimentation is not necessarily success, but learning from practice. The concept of prototype is relevant here. It changes expectations of performance and permanence of public services, given the signal of early-stage development and ongoing learning. Prototypes not only welcome feedback but proactively encourage challenges and critique from the public, potential users, colleagues, partners, experts, and other relevant actors. In this way imperfection becomes a legitimate and even expected part of the processes devoted to experimental polices.

5.3 Rethinking the Practice of Policy by Designing New Services

In the second part of the chapter, the Campus Sostenibile project is discussed as an example of a long-term, design-led project which aimed at developing new solutions for some of the societal problems that affected the Città Studi Milano neighborhood. In addition the project exemplifies a long phase of experimentation with a neighborhood-based coproduction network and the process of interaction between the network and the Municipality of Milan, thanks to the project's intermediation.

Campus Sostenibile was initiated to explore how to set up in a university a platform that could facilitate behavior transformation toward sustainability and that would be open to a neighborhood in the city of Milan. Fundamental strategic decisions were made by the management of the Politecnico di Milano in the adoption of a long-term perspective that could go beyond single projects, to lead a very transformative approach to participation and people's behavior in the campus as well as in the local communities of the neighborhood. At the beginning the project was not intended to be an instrument of transformation and urban regeneration. It was supposed to be a platform for supporting a dialogue between the people in the campus and the people in the neighborhood.

In the rest of the chapter, details on the overall project, from the initial idea to the reconfiguration of Campus Sostenibile as a design-driven living lab, are discussed. Next, the initial idea, motivation, and core solution and how they have been sustained through a strategy of synergies with different scale projects are described. They are followed by a series of design intervention descriptions with details on performance, scales, and funding schemes. Finally, lessons learned with respect to design as complex participatory practices are summarized.

5.3.1 A Story of a Complex Participatory Design Process Toward a Transformation of a Neighborhood in Milan

The Campus Sostenibile project tells the story of a complex participatory design process (Deserti and Rizzo; 2014) as the continuous open innovation strategy implemented by a public actor experimenting with open innovation in urban context. The elements that this story put together are a university, namely, the Politecnico di Milano (which is located in one of the largest neighborhoods in Milan, Città Studi); the citizens who live in the neighborhood; the university communities (students, professors, and administrative staff); and a series of private and public actors that operate in that context, the Peripheria European project (no. 271015) funded under the CIP-IP program for the development of the smart city paradigm.

When Campus Sostenibile started, the idea that the university could become the engine of the neighborhood transformation was not clearly declared. In fact it was the Peripheria project approach (Marsh 2013) that created the conditions to set up and generate organizational changes toward sustainability. It did so by exploiting open innovation as the most suitable form of collaboration between the university and the Città Studi neighborhood.

In this context, participatory design was introduced as a way of envisioning possible future solutions, by creating strong connections with the network of stakeholders belonging to the neighborhood and establishing a long-term engagement with local communities, which leads to the emergence of new practices and new opportunities for all.

The final result of this process has been the implementation of an intangible infrastructure, which can be also defined as an urban living lab (Concilio et al. 2012), in which local stakeholders continuously co-design and coproduce solutions to address situated challenges in the neighborhood. Designers work with stakeholders to identify the emerging needs and to create digital and physical platforms that can enable participation and coproduction, being open to different project development directions and perusing the sustainability of the designed solutions.

The case describes the peculiar conditions and resources of the local communities engaged in this long-term experiment at the beginning and at the end. The challenge is to provide evidence of what can be done beyond the classical co-design exercises, activities, and tools, with a twofold aim: (i) addressing the context's problems and (ii) establishing a long-lasting strategy of innovation for that context.

5.3.2 *Campus Sostenibile Program: The Core Solution and Its Further Development*

The idea of the Campus Sostenibile project started in October 2010 as part of the strategic vision of the new management of the Politecnico di Milano. The initiative announced in March 2010, during the opening of the academic year, initially aimed to improve the level of sustainability of the university, i.e., the sustainability of behaviors of the institution and the behavior of its related communities.

The program started from the most ancient campus of the university, the Leonardo Campus, located in the center of a large city neighborhood, Città Studi.

The first year of the program was spent designing new policies, rules, and services to encourage the people who were using the campus to change their behaviors and act in a more sustainable way with regard to energy consumption and building use in particular (two of the main issues). In 2010 the Peripheria European project had been just launched, with the aim of applying co-design methodologies to the design of smart services that were to be experimented in five different European cities. The project heavily relied on a design-thinking approach as the methodology to be applied in the design of smart services. The Politecnico di Milano, with the departments of Design and of Urban Planning, was included in the project as the partner that would provide the skills needed in the project and that would lead the co-design activities in the five city contexts.

At the end of 2010, the management decided to open the Campus Sostenibile to the support of its internal research community (engineers, designers, and architects). The idea was that many researchers in different fields are conducting activities that could be useful for the Institution for which they work as well, because they provide solutions, ideas, and knowledge that can been applied to achieve the objectives of the Campus Sostenibile program.

The group of researchers that was working for Peripheria immediately decided to join as an active member of the Campus Sostenibile program. The fact that the Campus Leonardo da Vinci was located in the Città Studi neighborhood was a great occasion for Peripheria to experiment co-design in a specific element of the city of Milan.

The Peripheria research team proposed to try a series of small-scale experiments, to open the scientific knowledge of the university to the citizens of that neighborhood, in order to start co-design processes to address some of the context's problems.

Relying on this main assumption, the goal perused with the synergy between Campus Sostenibile and the Peripheria project became to transform the campus into an urban experimental place where scientific knowledge is coproduced. Thanks to this vision, Campus Sostenibile planned to open the scientific world to a wider community in order to transform small-scale projects into occasions for experimenting the production of scientific knowledge as a large public collective experience. The collaboration started with the general idea of renovating and innovating the area of the Leonardo Campus in a sustainable perspective.

The original feature of the Peripheria project, compared to other smart city projects that the European Commission was funding in 2010, was the involvement of specific competences in urban planning and design for the conception of new, people-centered services, which would also represent "smartness" as the city's ability to develop solutions in line with the citizens' needs and desires. Here the Politecnico's competences on service design and urban planning found a new and tested common ground for research by further developing the idea of collaborative services (Baek et al. 2010; Prahalad and Krishnan 2008; von Hippel et al. 2011), because in urban contexts these services are the results of co-design and coproduction initiatives. From the Peripheria project, a new vision on the role that Politecnico could have as a public agency that operates within a city was raised in Campus Sostenibile as well. Città Studi neighborhood became an area for conducting small-scale collaboration experiments between the university and the other actors of the neighborhood, mediated by co-design techniques and with the aim of solving real problems.

By the time Peripheria started to work in the Città Studi neighborhood, the Campus Sostenibile program was already under way. However, it was experiencing difficulties in establishing channels to communicate and interact with the external environment, as well as with the internal communities, because it lacked a strategy and an approach for stakeholder and community involvement.

The first action that the Peripheria team decided on was the creation of an official table around which to meet stakeholders from the neighborhood. At the beginning of 2011, a "City Table" for Peripheria and Campus Sostenibile was opened, led by the researchers from the Design and Urban Planning departments, with representatives from the neighborhood (associations, schools, municipality). The table met on a weekly basis to discuss the problems of the neighborhood. It was the first time the Politecnico had opened its doors to the local outside communities.

After two exploratory workshops, the involved stakeholders decided to focus the table's attention on the Leonardo da Vinci Square.

This is a square that for many reasons has always been a kind of contradiction for the neighborhood. It is located just in front of the Rectorate of the Politecnico, but it is a public area under the municipality's responsibility. It is cut in the middle by a street where cars can cross. At the time Peripheria started, half of the square was a parking area for Politecnico staff, and the rest was a free area to walk in, located right in front a primary school. In the night the square is not under surveillance and it is considered a dangerous site. Many buildings overlook the square and the people that live there complain about the fact that the square is a park in the day and an empty and dangerous place in the night.

It was evident that the square was a sensitive issue for many actors of the neighborhood. The Politecnico was interested in changing the bad perception that citizens had of the university. Using the square as a park was perceived, by the other actors, as an abuse which the neighborhood was not prepared to put up with. The municipality and the inhabitants around the square were interested in transforming it into a safe place. The primary school on the other side of the square was interested in using it as an open area for children to play safely in. From the first of the table meetings, one topic appeared to work well to implement the synergy between the Campus Sostenibile program and Peripheria project, i.e., the transformation of Piazza Leonardo da Vinci. With this goal in mind, an initial network of stakeholders started to co-design in order to deal with a first challenge, which meant exploring the campus of the Politecnico di Milano as an experimental context for socio-spatial transformation.

The initial stakeholder network then started to disseminate the challenge and to design a series of events and actions with the aim of helping interested groups to emerge and work on the challenge, capturing and supporting emerging processes of alignment at different levels around the co-designed solutions to be coproduced in the real context.

The interventions required mobilizing a number of stakeholders and asking them to invest time and resources. This implied the expectations were often quite high, leading to a number of issues related to the actual results of each prototype, with failures being difficult to accept and manage. At the end of July 2011, three different networks of stakeholders were established with the aim of working on the challenge and developing design projects to answer it.

The strategy of developing a long-term engagement progressed through the design of service prototypes that put the square at the center of the challenge's solution. The network emerging around the Piazza Leonardo da Vinci theme was the most active. This was mainly due to the value assigned by many citizens, organizations, school, and institutions to this space. Despite this wide and deep interest, this network was the most risky because it was becoming more and more a sort of test for the municipality, which had to accept a new mode for making decisions, dealing with different and diverse visions, dialoguing with new actors, and accepting the Politecnico as a strong interlocutor that was not prepared to have the role of technical consultant as its only role.

During 2012 many ideas were produced around the topic of how to transform the square. Two stakeholders' networks emerged around two main projects:

- The transformation of the Leonardo da Vinci Square into a pedestrian area for children to play in and for students, but also as a venue to organize cultural events for the citizens
- A long-term project that would catalyze sport events for adults and for children in the Leonardo da Vinci Square

The Peripheria project ended in May 2013. In June 2013 the first Leonardo da Vinci sport program was launched, and in January 2014 the Politecnico decided to transform the part of the square under its responsibility from a park into a garden for the students (many Polimi staff are still complaining against this decision, claiming they have nowhere to park their cars). In spring 2014 the municipality experimented for 2 weeks during which the square was closed to cars and became a place for sport.

The Leonardo da Vinci Square networks described above have been the most relevant emerging from the Campus Sostenibile program.

At the moment a clear success in reaching the challenge objective cannot be announced, but the fact that the municipality accepted a different temporary mode for decision-making and the fact that it accepted to "test" some public functions in the square area before implementing any predetermined transformation are good signals.

Currently Campus Sostenibile has applied for the European Network of Living Labs (ENOLL) call for applications. The networks of stakeholders around the square under Politecnico leadership are continuing its action of continuous experimentations, i.e., new service prototypes, which can manage a smart use of the square from private citizens (Stick Around app) or that can implement mutual help processes between students and people from the neighborhood, have been developed or are under experimentation.

5.4 Case Discussion

Campus Sostenibile as a design-led project has focused more and more on building alliances among one leading partner (the Politecnico) and the internal stakeholders, the citizens, the representatives of the public sector in the neighborhood (the school, the municipality), and the representatives of the private sector (small shops, bars, and restaurant) with the aim of having an impact on the processes of decision-making and transformation for the Leonardo da Vinci Square. From this point of view, the first elements of discussion that the case allows to focus on are the specific characteristics of its initial configuration and the vision behind the processes of alignment that the project implemented.

The case of Campus Sostenibile represents a project in which one of the actors involved takes the leadership of guiding the design process, by envisioning a design project as an enabling platform for interplay between bottom-up experiments and top-down policy-making and regulation frameworks. The second element of the case on which to reflect is the nature of the process of building infrastructure. This process has also been discussed by Pell Ehn and his colleagues in many papers (Björgvinsson et al. 2012; Hillgren et al. 2011; Ehn 2008) as the process through which design helps to build linkages and supports small-scale initiatives to become connected (quotation), but also in the meaning of Manzini and Rizzo (2011) who conceive infrastructure also as the process of designing a design project to set the precondition in which to experiment with policy and people needs. "Infrastructures" basically mean that Campus Sostenibile cultivated long-term working relationships with diverse actors and slowly built a stable designing network that can change configuration with respect to the specificity of the challenges faced, the interests and needs of the different stakeholders, the constraints, as well as the affordances that the socioeconomic and regulation framework imposes and offers. Thanks to this long-term perspective, Campus Sostenibile built trust among diverse stakeholders, supported mutual learning, slowly gained the authorities' attention, and worked on a more systemic level.

The starting point for Campus Sostenibile has been quite different from the projects mentioned above. La 27e Région had several politicians sitting on their board, and their friendly hacking is commissioned by the public sector and supported by formal agreements that give them a mandate to work inside these organizations. Malmo Living Lab emerges, by contrast, from the fact that it started and rooted its initial work in local communities. Dott07 and Dott09 seem to represent attempts of a strong player like the Design Council to push design under the eyes of policy-makers and the public sector as the new approach to innovation.

Considered together, these projects can be read as a "framework program" (Manzini and Rizzo 2011) for cities, a large supporting structure that could move local cases, experiments, and projects out of isolation and increase their capacity to impact on the development of a new vision for cities.

When contradictions emerge between bottom-up and top-down in framework projects, processes of alignment start with the aim of producing a possible change in the bigger picture, by trying to modify regulations, work procedures and cultures, public policy, and indicators of project success (Deserti and Rizzo 2015).

Framework design projects recognize that there is a need for a more permissive innovation culture in the public sector and in policy-making, so that stakeholders can be allowed to experiment (and even to fail). To support these processes, they use the concept of prototyping quite extensively. But at the same time framework projects also recognize the value of discussing how regulations could be stretched and how things can be done without breaking any regulations or laws. To make this possible, framework projects develop broader vision and scenarios in which to discuss policy and through which inform policy decision-making.

All considered, the framework project reveals a model of conduction, here presented as a re-elaboration of a first model discussed in Manzini and Rizzo (2011), which reorganizes design activities in two larger phases.

What emerges here as original compared to the first version of the model is the idea of complex participatory design processes as the experimentations of coproduction networks along three implementation phases: infrastructuring, experimenting, and strengthening. Listed below are the phases of the model:

1. *Analyzing*. The exploration and mapping of existing solutions and initiatives oriented toward the inspiration of new solutions or systems of solutions. It includes the identification of a consistent design opportunity for a competitive and innovative solution.
2. *Envisioning*. The development of scenarios, visions, and proposals, used both to define the overall directions to take and to stimulate and align the actors and stakeholders in the development process.
3. *Designing*. The development of the solution through the adoption of participatory design tools supporting interaction and convergence among the involved parties.
4. *Communicating*. The development of presentations, visualizations, communication tools, and actions to inform about the solution before, during, and after its development, with different aims such as convincing potential actors to join or sponsor the initiative, create consensus, foster the adoption of the solution, etc.

5. *Infrastructuring*. The development of digital platforms, toolkits, and other supporting tools and actions (such as knowledge transfer initiatives), to enable the new network of actors to carry on the development process by themselves.
6. *Experimenting*. The solution experimentation in local scale and small scale, including the assessment and the testing of the involved actors' network, to give feedbacks for the assessment of the new idea.
7. *Strengthening*. The activities oriented toward organizing synergies and multiplication effects among different single projects and different elements of the same project.

The model suggests that the design phases (analyzing, envisioning, designing, communicating) are usually followed by a long-term period of experimentation (small-scale experiments) that aim to infrastructure the context of the project through the institutionalization of partnerships that strengthen and deliver sustainable solutions.

5.5 Conclusions

This chapter has discussed a new wave of design projects that are active in cities and territories. They aim to face new societal challenges through a situated approach and extensively applying design culture to experiment new networks and partnerships as infrastructures of a new generation of services that are much closer to society's needs.

Until now these design projects have been presented as top-down or bottom-up initiatives. Actions "from the top" are capable of generating large social transformation, e.g., the experience of La 27e Région, whereas actions "from the bottom" give rise to a multiplicity of local changes, such as the foundation of the Malmo Living Lab in Sweden or Dott07 and Dott09 in the United Kingdom. However, a closer observation indicates that this new generation of design projects, both in its starting move and in its long-term existence, often depends on more complex interactions between very diverse initiatives, where the ones undertaken directly by the people concerned (bottom-up) are often supported by different kinds of intervention by institutions, civic organizations, or companies (top-down). The chapter has discussed these as complex participatory design processes.

Considering these design projects from the perspective of design culture, three main common characteristics can be observed: (1) they aim at provoking transformations at some territorial scales (neighborhood, squares, streets, cities, regions) in order to address context-dependent societal challenges, (2) they share the explicit goal of achieving their objectives by activating networks and partnerships of diverse actors (from citizens to private and public stakeholders) in designing and producing solutions, (3) they have been started and are driven by the explicit intent to investigate design extent and potentialities in dealing with cities; and (4) all of them can be described as infrastructure that permits the interplay between small-scale initiatives and a broader vision for addressing city societal challenges. From this point of view, framework projects can be described as (Manzini and Rizzo 2011):

> 1. Highly dynamic processes: they include linear co-design processes and consensus building methodologies (i.e. the most traditional view on participatory design), but they can go far beyond them, becoming complex, interconnected but often-contradictory processes. 2. Creative and proactive activities, where the designers' role includes the role of mediator (between different interests) and facilitator (of other participants' ideas and initiatives), but involves more skills and, most importantly, it includes the designers' specificity in terms of creativity and design knowledge (to conceive and realize design initiatives and their correspondent design devices). 3. Complex co-design activities that, to be promoted, sustained and oriented, call for prototypes, mock-ups, design games, models, sketches and other materials: a set of dedicated and designed artifacts.

But from this point of view, framework projects can be described as intermediate playgrounds in which top-down policies leave bottom-up initiatives the room and the time to experiment with failure and success by taking advantage of a long-term learning process (Deserti and Rizzo 2015).

Each small-scale experiment that moves from the stage of social invention and working prototype and evolves to become a robust and replicable solution asks for a positive interplay among bottom-up initiatives (by grassroots associations), peer-to-peer exchanges (between similar initiatives), and top-down interventions (by local authorities, sensible businesses, other nonprofit associations). A key feature of these new design projects is the larger sphere of intervention that comes about creating a new, productive, quasi-institutional environment, which is held up by various actors, different power relations, and interconnected spaces of meaning and interpretation. These environments challenge the type of role public authorities have. Here the public state recognizes itself as one knowledgeable actor among many and therefore deliberately seeks to draw broadly on the knowledge and efforts of various actors of society as a whole.

Finally, it might sound like a revolution if design-led innovation were really to affect and impact governments around the world by complementary internal organizational cultures with design-based culture. As also noticed by Bason (2012) along this attempt, design has to rise to three main challenges:

- *The institutionalization of design environments within public organizations.*
  Despite the fact that many entities like design labs, centers, and teams have been established in many countries in Europe (United Kingdom, Denmark, Finland), public organization still struggles with the idea that design is an area of competences to be legitimated within public organizations and that it is also something to invest in, in order to ensure funding, anchor change in the organization, get management buy-in, and actually execute the new ideas and solutions into real prototypes.
- *The empowerment of public organizations through long-term processes of capacity building.*
  The introduction of design culture in the public sector is in its initial phases. Design methods and tools are still largely unknown to public institutions, and design knowledge is still far from having entered the public organizations at a

large scale, affecting their daily processes and their underpinned culture. Campus Sostenibile shows the possibility of building an intermediate playground, for a dynamic interaction between the operative and the strategic levels of organizational change. The project shifts the attention on how to obtain a wider impact through the introduction of new policies to design and experiment new ways of delivering services, using the experiments to assess the policies and to foster the change of the involved organizations (macro level).

- *The organizational change issues are actually unknown to most of the designers.* The cases show how the conception and delivery of the new services are bound to the creation of networks and partnerships which, in turn, requires the development of new policies. Some of the service design tools, such as the "actor mapping" and the "stakeholders' matrix," apparently put both feet in the field of organizational change without a sound understanding of its complexity.

References

Baek J, Manzini E, Rizzo F (2010) Sustainable collaborative services on the digital platform. In: Proceedings of design research society, DRS2010. DRS, Montreal

Bason C (2010) Leading public sector innovation: co-creating for a better society. Policy Press, Bristol

Bason C (2012) Designing co-production: discovering new business models for public services. In: Proceedings of DMI conference, DMI, Boston

Binder T, De Michelis G, Ehn P, Jacucci G, Linde P, Wagner I (2011) Design things. MIT Press, Cambridge

Björgvinsson E, Ehn P, Hillgren PA (2012) Design things and design thinking: contemporary participatory design challenges. Des Issues 28(3):101–116

Botero A, Saad-Sulonen J (2013) Peer-production in public services: emerging themes for design research and action. In: Manzini E, Staszowski E (eds) Public and collaborative: exploring the intersection of design, social innovation and public policy. DESIS Network, Milano

Brown T (2009) Change by design: how design thinking transforms organizations and inspires innovation. Harper Business, New York

Brown T, Wyatt J (2010) Design thinking for social innovation. Stanford Soc Innov Rev 9(1). http://www.ssireview.org. Accessed 5 May 2015

Burns C, Cottam H, Vanstone C, Winhall J (2006) Transformation design. RED paper 0.2. Design Council, London

Christiansen J, Bunt L (2012) Innovation in policy: allowing for creativity, social complexity and uncertainty in public governance. http://www.nesta.org.uk/publications/assets/.features/innovation_in_policy. Accessed 5 May 2015

Concilio G, De Bonis L, Marsh J, Trapani F (2012) Urban smartness: perspectives arising in the periphéria project. J Knowl Econ, Springer

Deserti A, Rizzo F (2014) Design and the cultures of enterprises. Des Issues 30(1):36–56

Deserti A, Rizzo F (2015) Design and organisational change in public sector. Design Mana J 9(1)

Design Council (2008) Annual report. Design Council, London

Design Council (2012) Annual report. Design Council, London

DiSalvo C (2012) Adversarial design. MIT Press, Cambridge

Ehn P (2008) Participation in design things. In: Paper presented at proceedings of the 10th biennial participatory design conference, ACM, New York

HM Government (2012) The civil service reform plan. http://www.civilservice.gov.uk/wp-content/uploads/2012/06/Civil-Service-Reform-Plan-acc-Thefinal.pdf. Accessed 5 May 2015
Hamdi N (2004) Small change: about the art of practice and the limits of planning in cities. Earthscan, London
Hermant-de-Callataÿ C, Svanfeldt C (2011) Cities of tomorrow: challenges, visions, ways forward. http://ec.europa.eu/regional_policy/index.cfm/en/information/publications/reports/2011/cities-of-tomorrow-challenges-visions-ways-forward. Accessed 5 May 2015
Hillgren P-A, Seravalli A, Emilson A (2011) Prototyping and infrastructuring in design for social innovation. CoDesign 7(3–4):169–183
Jégou F, Vincent S, Thévenet R, Lochard A (2013) Friendly hacking into the public sector: co-creating public policies within regional governments. In: Proceedings of boundary-crossing conference on co-design in innovation, Helsinki, Aalto University. http://www.slideshare.net/27eregion/friendly-hacking-into-public-sector. Accessed 5 May 2015
Julier G (2006) From visual culture to design culture. Des Issues 22(1):70–92
Julier G (2013) From design culture to design activism. Des Cult 6(2):216–236
Latour B (2005) Reassembling the social: an introduction to actor-network-theory. Oxford University Press, Oxford
Latour B (2010) An attempt at a compositionist manifesto. New Lit Hist 41:471–490
Manzini E, Rizzo F (2011) Small projects/large changes: participatory design as an open participated process. CoDesign 7(3-4):199–215
Manzini E, Rizzo F (2012) The SEE project: a cases based study to investigate the role of design in social innovation initiatives for smart cities. In: Campagna M, De Montis A, Isola F, Lai S, Pira C, Zoppi C (eds) Planning support tools: policy analysis, implementation and evaluation. Proceedings of the seventh international conference on informatics and urban and regional planning INPUT 2012. Franco Angeli, Milano
Manzini E, Staszowski E (2013) Introduction. In: Manzini E, Staszowski E (eds) Public and collaborative: exploring the intersection of design, social innovation and public policy. DESIS Network, Milano
Marsh J (2013) The peripheria CookBook. DG Connect, Brussels
Mau B (2004) Massive change. Phaidon, Nachdruck
Mulgan G, Leadbeater C (2013) Systems innovation. Nesta Discussion Paper, http://www.nesta.org.uk/publications/assets/features/systems_innovation_discussion_paper. Accessed 5 May 2015
Murray R, Caulier-Grice J, Mulgan G (2010) The open book of social innovation. Nesta, London
Norman W (2012) Adapting to change: the role of community resilience. The Young Foundation, London
Prahalad CK, Krishnan MS (2008) The new age of innovation: driving co-created value through global networks. McGraw-Hill, New York
Staszowski E, Brown S, Winter B (2013) Reflections on designing for social innovation in the public sector: a case study in New York city. In: Manzini E, Staszowski E (eds) Public and collaborative: exploring the intersection of design, social innovation and public policy. DESIS Network, Milano
Tackara J (2007) Wouldn't it be great if... designs of the time manual. Design Council, London
UK Design Commission (2013) Restarting Britain 2. Design and the public services. UK Design Commission, London
von Hippel E, Ogawa S, de Jong JPJ (2011) The age of the consumer-innovator. MIT Sloan Manag Rev 53(1):27–35

Chapter 6
Spatial Planning and Design Citizenship

Guy Julier

Abstract The application of design within city planning and governance has hitherto taken various, related guises. It is to be found in the renovation of streetscapes for urban regeneration, the development of place-brands, through policies that promote the 'creative city' and within cultural planning. Each of these represents relatively 'top-down' approaches. More recently, design has been drawn closer into public sector innovation processes in some instances. This may be read as part of a shift in public policymaking from New Public Management to network governance. It also involves a move from design as an objectoidal outcome to design as process. It includes a more person-centred, results-driven approach that pragmatically looks to the most efficient outcome in terms of public sector service delivery. This challenges traditional bureaucracies and spatial arrangements for it invariably transcends their usual hierarchies and boundaries. Key drivers here have been the multiple challenges that have been set through austerity agenda in public policy. New alliances and hybrid possibilities for connecting public administrations and citizens may emerge. This chapter traces some key tendencies and examples here, in order to critically analyse the possibilities and limitations of a concept of design citizenship.

Keywords Design • Citizenship • Governance • Planning • Austerity

6.1 Introduction

Spatial planning and design citizenship exist at opposite extremes in the conceptualisation, decision-making and lived experience of territory. The following are complex questions that command attention across the management and social and political sciences but also in design-related fields that include social innovation studies, urban design, strategic and service design, design in public sector innovation, architecture

G. Julier (✉)
Victoria and Albert Museum, London, UK

University of Brighton, Brighton, UK
e-mail: g.julier@vam.ac.uk

G. Concilio, F. Rizzo (eds.), *Human Smart Cities*, Urban and Landscape Perspectives 19,
DOI 10.1007/978-3-319-33024-2_6

and the built environment: how cities are assumed to function and how its functions are distributed, what and whose interests are foregrounded and in what patterns of governance and how citizens see themselves and their roles in the urban fabric. A complex matrix of academic and professional concerns is at play here.

Spatial planning infers a top-down viewpoint, both in cognitive and governance terms. In the orchestration of resources and potentialities, spatial planners stereotypically look down at maps and statistics, arranging economic, social, cultural, technological and topographical data into stratagems that reflect various policy priorities. Their work is somewhat militaristic in its tactics of spatial reinforcement. However, where it departs from this metaphor is that it normally takes place within clear demarcated boundaries, be those of a municipality, a local authority, a region or a nation.

Design citizenship does not exist as an 'official' term. It does not appear as a 'core aim' in 'city visions'; nor is it taught in schools or universities. It might be taken as a way by which the material and immaterial features of everyday life, and the processes that produce these, make and unmake citizens (Weber 2010: 11). How does design include and exclude individuals and groups in their membership of society? How is design sometimes used to iron out difference or dissent in what it means to be a citizen (Fortier 2010)? Design citizenship might be more active and inclusive, however. How do people participate in the shaping of their surroundings and the multiple interactions that constitute day-to-day living or, indeed, make it extraordinary? Whether we are talking about passive or active relationships here, what is embedded in this idea of design citizenship is that its point of departure is more intimate, less easily defined or schematically represented: it is certainly not bound by the same cartographic regimes of spatial planning.

Aesthetic markers have been used to get people to support or to enforce oppressive ideological systems. Examples of these might include the wearing of youth organisation uniforms, mass pageants, commemorative statues that fill out every key public space, blanket advertising campaigns, the saturation of mass media with football coverage or racially segregated public amenities. But design citizenship can also be heterogeneous in its performance and self-representation, multiple in its politics and varied in its economic objectives. It all depends on who and where we are talking about.

In this chapter, I take the broad ideological context for exploring the relationships of spatial planning and design citizenship to be that of neoliberal democracies. This is double-edged, however. On the one hand, its promise is taken to be of agency, plurality and differentiation. On the other, hegemony may continue to operate. Briefly, this can be understood in terms of the dispersion of power across a discursive field, wherein all objects are constituted as objects of discourse. Following Laclau and Mouffe (2001: 109), these objects 'pierce the entire material density of the multifarious institutions, rituals and practices through which a discursive formation is structured'. By contrast with spatial planning, design citizenship begins at the experiential front-line of everyday practices – places where the friction with those objects of discourse takes place in palpable terms. Design citizenship is about being and acting in various modes where the artefactual field (both material and immaterial) is understood to be active and acted on. If, then, these are followed up on, its actors can work backwards to understand or, even, change the bigger systems, bureaucracies or regulations that frame these. In other words, design citizenship may invoke a participatory, bottom-up or co-creative process.

The gap between this version of design citizenship and the top-down, regulationist, bounded and bureaucratised practices of spatial planning is mediated by 'soft spaces' and 'network governance'. Soft spaces disrupt the order of spatial planning in that they cut across official or traditional territorial domains. They are pragmatic inventions that work with or outside these. They are in-between, overlapping and fuzzy in their boundaries. As such, they can evade or disturb official and traditional administrative structures as well, by calling on multiple-agency, cross-sectoral and layered collaboration between interests. Given this multiplicity of actors who are involved, the binding systems of governance that are employed are softened. Instead of rules and regulations, we have guidelines, understandings and agreements. Arriving at these is also complex, involving the garnering of expertise and viewpoints from a range of stakeholders. In short, this is, supposedly, where network governance takes place.

If we take the active involvement of stakeholders in network governance also to be, in part, design citizenship, then it remains for us to look more carefully at how designs and design processes are functioning here. This chapter explores in more detail the use of design in shifts between spatial planning and soft spaces of governance, considering through examples what versions of design citizenship are available.

6.2 Design in the City

In national and regional governments, the rise of design has been understood and promoted largely in terms of its role in commercial or industrial sectors. Successive government-funded organisations (almost universally termed 'Design Councils') dedicated to the promotion of design have placed emphasis on its role in fashioning competitive and differentiating products for the global marketplace, an assumption that is reflected in related academic literature (e.g. Sung et al. 2007). Governmental creative industries policies have gathered pace across the globe from the late 1990s and within these designs have largely been measured in terms of its turnover and contribution to GDP, its export capabilities, numbers employed and other indicators that demonstrate its economic standing and strength. Its recurrent appearance in statistical reports regarding the wider creative economy (e.g. UNCTAD 2010; DCMS 2014; FIRJAN 2014) underlines its commercial status within global trade networks.

Within this narrative, the design industry's growth and intensification in the service of these are to be found in different formats of its spatial clustering.

Examples of this would include the establishment of corporate design centres of global brands such as BenQ in Taipei or Ford, LG, Nokia, Samsonite or Sony in London. These exploit both the resource of available creative workers and the notion that multicultural, cosmopolitan cities can somehow act as localised consumer test-beds that reflect the global marketplace. These exploit the cultural, social and knowledge capital at hand, fostering a certain urban habitus within its corporate design culture that is subsequently reproduced as a promotional device through its marketing and brand image (e.g. Temporal 2006: Chp. 4; Design Week 2009).

This clustering and intensification notion extends into spatial planning policies that run from the encouragement and support of creative quarters at the intra-urban level (Evans 2009) to larger scales of regional policy (Gordon and McCann 2000). A stark example of this is the denomination of Shenzhen as 'design city' within China's policy of Special Economic Zones (Zhang 2008; Keane 2013) which, in turn, supports the wider manufacturing region of the Pearl River Delta. This policy builds the area's reputational assets of innovation (Sum 2010). Other examples include Antwerp, Glasgow, Lisbon, Montreal, Saint-Etienne, Stockholm and Times Square, New York, who coalesced as a network to debate the 'cities' different strategies for 'positioning and growth through design' (Lacroix 2005: 15). Since 2006 the 'UNESCO City of Design' programme as part of UNESCO's 'Creative Cities Network' initiative has bestowed the title on various global cities including Seoul, Buenos Aires, Cape Town, Dundee, Shenzhen and Berlin.

The imbrication of design objects and design economies into spatial and economic planning begins to be revealed here. In answering its own question 'Why Cities?', UNESCO provides the standard answers that cities provide the 'entire range of cultural actors necessary', are 'breeding grounds for creative clusters', carry strong public/private partnerships that 'unlock the creative entrepreneurial potential' and are small enough to consolidate cultural industries but are also 'gateways to international markets' (UNESCO 2015). A global circulation of what it means to be a design city, framed by neoliberal notions of competitivity, creativity and entrepreneurialism, is at work here. Further, there is no getting away from the narrative that design is a wholesomely urban practice. We read of 'design cities'. A 'design town', unlike a book town (Seaton 1999), sounds absurd.

But to get beyond the broad 'design city' rhetoric, there are four dominant strategies through which cities have exploited design objects and economies in order to gain competitive advantage. These are the energetic deployment of urban design to revive the cityscape, the use of place-branding to strengthen and build enthusiasm for its internal and external image, the encouragement of creative industries particularly through clustering and the overlaying of cultural planning that boosts reputation in social and cultural capital. Each has been challenged for different reasons, and while plenty of examples of these continue to be applied, we may also view them historically in terms of an evolution towards other, more recent, approaches. Knowing their limitations helps in building a more rounded view of design citizenship.

6.3 Design in Top-Down Urban Development Strategies

The concept of design-led urban regeneration or development places the fashioning of aesthetic resources and markers at the centre of spatial planning. This is in their productive capabilities and in the social milieux that work within them (Bell and Jayne 2003). This contrasts with, for instance, strategies such as the provision of direct subsidies to industry through tax relief or indirectly through infrastructure investments. From the late 1990s, the strategic use of design within urban regeneration was marked by the well-known tactics of urban design, place-branding, creative industries development or cultural planning. Each of these has their limitations.

Since the use of urban design in regeneration processes in Barcelona from the 1980s was launched, the provision of new pedestrianised areas, plazas and squares has grown as a way of stabilising civic spaces, showing the city to be up to date and attracting inward investment. A criticism of this approach has been in the design processes that have been employed. It has often involved broad brushwork in applying codified design systems without particular attention being paid to local needs and, sometimes, with low build quality (Urban Task Force 2005; Julier 2009). While praised as being a tool for bringing dignity into the public realm and, indeed, helping to foster a degree of participation in civic life (Calavita and Ferrer 2000), it has also been criticised as a 'quick fix' that doesn't necessarily deal with underlying social and economic challenges (Delgado 2007). In addition, it has been seen as working hand in hand with gentrification processes to privilege the ambitions of property developers (Boddy and Parkinson 2004; McCarthy 2007; Syrett and North 2008; Kauko 2010).

Second, place-branding may be similarly fragile. This draws from the idea that a location can be dressed as a product (Anholt 2002; Papadopoulos and Heslop 2002; Bennett and Savani 2003). Taking a 'fast' approach to establishing and communicating identity, slogans, logos and other forms are mobilised to tell a story of a place (Evans 2003). A criticism of this method is that it invariably provides a layering over places that can ignore cultural difference within them while pandering to neoliberal notions of cosmopolitanism and enterprise. It takes place within assumptions with regard to inter-territorial competition for residents, visitors, jobs and finance (Storper 1997; Buck et al. 2005). It can lead to a conflict between the hard sell and the actual product.

Third, the use of creative industries as a mechanism for the economic, social and cultural development of cities has been most vigorously championed by Florida (2002). Many of his assertions have since been challenged, not least because of their circular logic (Peck 2005). There is an assumption here that in attracting the creative class, it somehow magically clusters together. In terms of design, this idea of 'clustering', already referred to above, is challenged by Reimer et al. (2008), who demonstrate that designers are more concerned to locate close to clients rather than each other. In other words, a 'creative city' requires infrastructure, pre-existing demand and resources. Creativity can't just be dropped in.

Fourth, cultural planning, with the creation of locations or events to draw inward investment and boost tourism, is also a well-discussed regeneration tactic (Evans 2002; Stevenson 2004; Brecknock 2006). In broad terms this leads to a 'cultural arms race' (Hassan et al. 2007), wherein places are constantly having to deliver new projects for the attention economy. For design, this has led, for example, to the exponential rise of design festivals, biennales or weeks. Not only does this become a homogenising force where similar formats are reproduced globally (Gonzaga 2012), but it also contributes to a view of design as ephemeral and short term in its impact as initiatives come and go in accordance with the annual cycle of the cultural calendar (Berglund 2013).

Design is embedded into each of these four approaches. These are invariably deployed in relation to each other by town, city or regional authorities under various forms of 'strategic vision'. However, they are usually developed and implemented by separate departments. Thus, usually, urban design is managed through planning and architecture departments, creative industries are fostered in economy or enterprise sections, cultural planning comes within leisure services and place-branding is the responsibility of marketing departments. As such, each of these is directed towards specific material outcomes, be they, for example, a paving scheme, a media and design hub, a theatre or concert hall or a branding scheme.

Here, a vision is turned into a strategy which is then moved through plans that are undertaken by departments, culminating in their materialisation. As such, it is in fact difficult to conceive this approach as being truly design *led*. Design is conceived here more as an *output* that may or may not lead on to other things (inward investment, attraction of skilled knowledge economy workers, stimulation of enterprise culture, etc.). Outcomes are therefore uncertain. It also conspires in processes of 'agglomeration boosterism' where pre-existing trajectories are built on (Haughton, et al. 2014); it contributes to 'quick fix' strategies by leveraging pre-existing strengths. A 'design city', for example, is constructed around an established infrastructure of professionals and their associations, institutions including its art and design schools, sites of designerly consumption such as shopping or gastronomic quarters and mediatory forms such as cultural bodies. In turn, this produces uneven economic geographies. Spaces outside and between such agglomerations are destined to become economic wastelands. The agglomerations themselves triumph, but also run the risk of overheating through, for instance, property rental rises, the outstripping of demand by supply of creative workers and the exhaustion of available markets.

Each of these orthodoxies has different and related limitations. In terms of citizenship, the expert knowledge of planners or policymakers is separated from the various 'holders' (Schmitter 2002) of their effects. Those who enact and carry these plans and policies – the creative workers, the cultural producers, the ordinary citizens and so on – are rarely, if it all, engaged in their shaping. Design as a deliberative act is largely separated from citizenship.

6.4 Producing Design Citizenship: Two Examples

Design can be entangled more closely with citizenship, however. To return to the example of Barcelona in the 1980s and 1990s, the strength of the city's revival through design came through the coincidence of all four strategies discussed in the last section within buoyant economic and political circumstances. Citywide, highly creative urban design schemes; a series of rebranding programmes that stressed its Mediterranean, modern and European qualities; an ascendant creative industries sector centred, in particular, on new fashion; and interior and furniture design and production each worked alongside major strides in cultural programming, not least in the run-up to the 1992 Barcelona Olympics. These should be read against a closely knit network that brought design professionals and associations, mediatory forms such as the retail outlets and designer bars as well as local press and broadcasting interests and city and regional government into very close orbit (Julier 2005: 878). We should regard this as coincident but not programmatic. There was no overall coordination of these. Instead, they emerged as a hegemonic process driven by several economic and political factors, not least the ambition to re-establish Catalan cultural interests within European and global frameworks of modernity.

Nonetheless, this intensification may be interpreted as a clear example of a design culture where the close interaction of the work of design and its production, mediation and consumption is evident and is driven by an overall future-directed cultural and economic logic. However, this existed within advantageous financial circumstances of sustained inward investment and growth. All four of the aforementioned strategies are capital intensive: they require high investment in urban schemes, buildings and/or visual communications, let alone the consultants' fees in fashioning these. Arguably, the limitations and inefficiencies exhibited in each of the above four strategies for design-led, urban development are mitigated by this. In a more challenging economic environment such as that experienced in many cities since the financial crash of 2008, their efficacy and, even, affordability can be questioned.

In terms of design citizenship, popular buy-in to and interest in these strategies were achieved through their multilevel, iterative mediations that coursed through the city. Locally produced designer furniture was reproduced through retail outlets, bars and restaurants and in the public realm (Narotzky 2007). Graphic campaigns were revealed and discussed through local TV and print media, deployed through the city and critiqued around the dinner table (Busquet 1993). Ultimately, though, this was about the objects and those who designed them. While popular debate abounded, design citizenship did not extend to a deeper enrolment of Barcelona's population into design processes or, at the very least, knowing what these entailed. Rather, design functioned within the city's discursive field by providing an important and highly visible material density through which issues of modernity, Europeaneity and Catalan identity were sedimented.

A lesser-known but much more programmatic example where there has been an attempt to revive a city through design policies has been in Kolding, a small city in Denmark (Julier and Leerberg 2014). While Kolding includes one of Denmark's two dedicated design schools, a university campus focused primarily on design-related studies and an international business academy, few of the city's 4500 students settle in Kolding, while their lecturers mostly commute from Copenhagen, Aarhus or Odense. With an extensive seaport and with good motorway links to northern Germany, the city is an important logistics hub. However, the global economic crisis from 2008 impacted negatively on this sector. Further, the building of a 62,000 m^2 shopping centre on the city's outskirts in the early 1990s, with 120 shops, contributed to the hollowing out of its centre. Shops with 'to let' signs flanked many of its streets while its night-time economy is markedly quiet.

In 2012 the Kolding Municipal Council unanimously adopted a new vision for the city and municipality: 'Together, we design a better life through entrepreneurship, social innovation and education'. In a subsequent, shorter version, this became 'Kolding – We design for life'. Design was to be at the centre of all city development activities within a 10-year programme. The detailing of this came through a number of participatory workshops with citizens and local government officers and leaders to create a series of specific strategies that carried the three themes of entrepreneurship, social innovation and education forwards. A fourth, of design and branding, was directed more internally in terms of raising local awareness of the programme. Part of this was operationalised through appointing 50 civic actors from local business, education, culture and community associations who were enrolled into the Design Network Kolding. These would be ambassadors for the vision and network by contributing to development and implementation of its initiatives.

In this example, there is a turn away from a conception of design as some kind of object that might lead to other benefits, to using design in terms of its processual potential. In other words, design is seen, in part, as a skill, attitude or, even, disposition that may be embedded into the organisation (Michlewski 2008). But there are two other important developments being made. Hitherto, the relationship of these elements to innovativeness and organisational performance has been examined in terms of commercial entities (Press and Cooper 2003; Boland and Collopy 2004). The Kolding example shows how these begin to be deployed within a municipal authority, but also with the ambition that they are carried through multiple channels in the civic life of the city. Thus, they are to course through business activities and its various forms of welfare provision and also act as a springboard for all levels of education and training. They are expected to be engaged among multi-agency, multi-actor and multilayered interests that go beyond the single organisation.

The second development that emerges here is in the co-dependency of the processual and the material. In thinking about organisations, Deserti and Rizzo (2014) pay attention to the interplay of these more abstract, generalisable competences of design knowledge, understanding and skill and to the context-specific features that include the structuring of organisations and the physical resources at hand.

They acknowledge the trial and error, prototyping kinds of action that go on here. By exclusively focusing on abstract qualities of competences, we are left with frictionless and idealised assets. Alternatively, by considering their entanglement with actual, situated and material qualities of context, something more experimental and open-ended appears to emerge.

The Barcelona example of the 1980s and 1990s provides a compelling example of the entrepreneurial city that networks many aspects of everyday life around design. An overarching design culture emerged but without a coordinated programme of design-led planning. Ultimately, design was the object rather than the process here. Contrastingly, in Kolding from 2012, something more programmatic that ties together the objectoidal and the processual and a number of levels has been at play.

In the historical gap between these two examples, one from the 1980s to the 1990s and the other beginning in 2012, there have been important developments made in how design may be used and practised – particularly in the public sector – while, at the same time, challenges to traditional spatial planning approaches have begun to open up more spaces for the former. Let us deal with developments in the public sector design first.

6.5 New Design Spaces in the Public Sector

So far, with the exception of the Kolding example, the role of design in urban development has been discussed in terms of an overarching view of the city and the material interventions that stimulate other outputs such as social or communitarian capital, a stronger culture of innovation, inward financial investment or growth in tourism. Design has been used in its more traditional format – that of shaping objects – in support of traditional, economically centred policy goals.

Over the past 15 years, increasing attention has also been given to a more complex role for design in the public sector that focuses on its role in developing and implementing policies and therefore delivering services. Notable here is a move towards 'network governance' where 'the role of the state is to steer action within complex social systems rather than control solely through hierarchy or market mechanisms' (Hartley 2005: 30). Service design, strategic design and design for social innovation become entangled in ideas of co-creation and participation. Two types of practice illustrate how design is foregrounded in both developing more complex, sophisticated approaches to design and embedding these into public sector systems of governance and social innovation.

Public sector innovation takes in many related disciplines and practices including social innovation, customer experience, social entrepreneurship and service design. The latter, service design, has emerged as a response to the growth of the service sector in postindustrial economies (Saco and Gonsalves 2008; Stickdorn and Schneider 2010). It is concerned with the arrangement of multiple artefacts that constitute a service environment (such as a combination of web, smart card and products) and their placing and order. A key component of the service design approach is foregrounding research into the user 'journey' or experience. As such, the *relations* and *exchanges* that go on within this come into view. Another field of practice that is of note is 'design thinking' which has found increasing prominence in mainstream business school teaching. This has a varied history (Kimbell 2011, 2012), but most recently, it has come to involve the iterative processes of user observation, visualisation, prototyping and analysis. Rather than being by predetermined business management routines, it infers a more exploratory or experimental approach through creatively trying out and testing ideas.

In both these cases, and many more related ones, we find hierarchical distinctions between various actors being challenged. In particular, the importation of design thinking into management and control systems moves away conceptually from the dominance pre-existing delivery structures to a more flexible and practical arrangements. This suggests a turn away from the privileging of the bureaucracies of service delivery to a more pragmatic attitude that consider what outcomes are required and working backwards from there to consider the configurations and interactions that are necessary to produce these.

Within design discourses, this recalls Manzini and Jégou's notion of 'results'-driven design wherein responses to problems are not predetermined in terms of design outcome; rather, the most effective and appropriate (particularly in environmental and social terms) response is sought (Manzini and Jégou 2005). This calls for a radical reconceptualisation of design's aims, processes and outcomes. By prioritising results over means to achieve these, design is immediately taken outside its sub-disciplinary structures of 'graphics', 'interior', 'product' and so on. It becomes media agnostic where the 'best tool for the job' prevails over predetermined, specialist approaches.

In local government discourses, resonance is to be found in a turn in public service delivery being reconfigured around 'outcome-based budgeting' (OBB) or 'outcome-based commissioning (OBC) (KPMG 2011; Law 2013). Here, instead of thinking organisationally and financially in terms of the operations of a service structure, OBB looks to what one wants to achieve at the user end. As such, it is very user centred in the emphasis it lays on the desired results of services (healthier citizens, cared-for elderly, literate children, for instance) and 'reverse engineering' from there in thinking about how best to achieve these in terms of what combination of organisations, departments and institutions can best (and often most cheaply) provide that solution.

In both the design and social innovation for sustainability motivations as expounded by Manzini and Jégou, or in the OBB or OBC thinking around local government service delivery, there is a change of emphasis away from administrative structures towards the *front-line* where the material culture of everyday decision-making and action comes into focus much more sharply. The assemblages and relationships that ensue are more fluid and heterogeneous as compared with a state bureaucratic model. They do not necessarily accord with customary administrative boundaries or policy processes to be found in planning. The result, then, are new territories of governance.

6.6 Design and Soft Spaces of Governance

The traditional administrative order of planning offices, economic development departments, city marketing bureau and so on carried responsibilities that were defined by clear spatial boundaries and, indeed, accountable to an identifiable body of voters. A local, metropolitan, regional or national government is elected to produce and implement policies that exist and are enacted within its respective territory. Conventionally, we might then think of various governmental departments made up of public employees whose jobs are to implement these, first by drawing up clear guidelines and secondly by following them. This is what we might call the 'hard' spaces of governance reflected in regulatory planning.

The obvious converse of this is in 'soft' spaces of governance that exist 'outside, alongside or in-between the formal statutory scales of government' (Haughton et al. 2013: 217). In planning terms these might include area masterplans, enterprise zones or multiregional growth strategies that don't fit neatly into the statutory spatial frameworks of government (Haughton et al. 2010: Chap. 2). These spaces emerge, according to Allmendinger and Haughton (2009, 2010), in three ways: one is in bottom-up functional planning where, typically, their end users are not necessarily concerned with 'official' scales and boundaries; another is where these spaces are produced 'top-down' through governmental decision-making; and the third is where they have a more instrumentalist role in delivering other urban development projects such as in the case of cross-border economic development zones. Olesen sees this tendency more loosely, arguing that often soft spaces emerge out of a meeting of top-down and bottom-up interests. He sees them as the result of '(i) new imaginations promoting new informal planning spaces located outside the formal planning system and formal scales of planning and (ii) new networked forms of governance seeking to work outside the rigidities of statutory planning' (Olesen 2012: 911).

This connection between soft spaces and network governance means that we needn't just view the concept as being solely territorial. The development of local government processes from New Public Management to network governance displays a gradual drift into this fuzzy area. From the 1980s, New Public Management approaches encompassed a shift to private business-style forms of public service delivery with efficiency targets and measurement, the pursuit of 'best value' and, thus, a move towards the outsourcing of functions to commercial and not-for-profit entities such as charities or voluntary organisations. Territorially, this mixes local service facilities with the extra- or non-territorial interests of national, multinational or other kinds of providers. In terms of motivations, again this is mixed.

Network governance emerged from the 2000s to take this a step further by distributing the knowledge and decision-making processes more widely. In other words, the different actors that are engaged around a particular public problematic are all engaged to varying levels in configuring their solutions. This involves a new kind of politics that is characterised by a 'decreased confidence in government as arbiter and deliverer of improved lives ... where the state has not so much shrunk as reformulated its rationale and role from being arbiter and provider of key forms of collective infrastructure, preferring instead narratives such as enabling, facilitating, guiding, coordinating, occasionally stimulating' (Haughton et al. 2013: 221). Soft governance therefore features guidelines more than rules. Part of the rationale here is that this allows room for small-scale innovations, adaptations or modifications, respecting the agency of stakeholders in the organisational field (Brandsen et al. 2006).

Design objects have two roles here. The first is in the creation of boundary objects that function as interlocutors between actors within systems of network governance. A simple example of this might be, as seen in the UK, the development and publication of urban design guidelines such as best-practice guidance, area statements or neighbourhood design guides. These are usually undertaken through collaborative processes between local government planning offices and various interest groups in civil society such as heritage organisations and neighbourhood associations, but are frequently promoted by urban designers and/or their professional associations (Julier 2009). Typically, they specify an overall 'palette' of public realm design features such as materials, treatments and detailing that should be used that 'fit in' and characterise a locality. As such, they provide a loose reference point of regulation around which citizens and planners coalesce. Design considerations are visualised as a kind of 'metadata' to coordinate and, hopefully, align interests. They are not strict regulatory documents; rather they function as a point of reference, particular in resolving disputes between, for example, a property developer, a planning department and local citizens.

A second tier is in the way that design tools are employed as part of a collaborative process of determining soft policy. These tools include the use of models of neighbourhoods in so-called 'Planning for Real' process where citizens add their local expertise and voice to thinking through priorities and configurations in the built environment. In policy planning, cards, Post-It Notes or models may be used in prototyping systems – such as crime reporting (Drew 2015) – in order to produce a speculative evidence base. By taking it beyond an abstracted flow chart or set of PowerPoint slides and materialising, at least in model form, the actual elements that make up a system, a wide constituency of interests can understand how their components fit together and where and how to engage with them. In turn, this allows greater fluidity between participants in the development process so that the 'expert knowledge' of professionals doesn't dominate the agenda.

In both these instances, artefacts are not the end result of a design process but form part of the public materialisation of the process itself. They help to make decision-making procedures public. They are also contingent in that the understandings, actions and material outcomes that stem from them are not inscribed in or through them. They act as pointers or exploratory devices rather than definitive statements on how things should be. In this respect, as prototypes they enter into a temporal frame. They carry a futurity as 'things-that-are-not-quite-objects-yet' (Jiménez 2014: 383). At the same time, their appeal to the temporal imagination still takes place in the spatial context of urban development. Some of their roots are to be found in 1970s utopian discourses of do-it-yourself movements or cybernetic philosophy (Turner 2006). But its situatedness means that the prototype doesn't necessarily make for an 'everywhere'. Whether it is created to address procedures at national, regional or local levels, these are still influenced by a range of specific contextual factors, from legal considerations to the availability of particular resources, to customs or to cultural practices. Needless to say, given the fact that soft spaces of governance work across traditional structures, these are rarely going to be unified and are always complex.

6.7 Design Citizenship: Fors and Againsts

What does design citizenship have to offer? If we are to take it to mean the active involvement of people and institutions in the arrangement and shaping of urban resources and practices, rather than it being a kind of semi-passive, compliant adherence to prescribed functions, then it proposes the following:

- A pragmatic approach that puts delivery mechanisms at the service of solutions rather than the other way around
- A turn from design as solely the objectoidal outcome of privately rendered activities to it becoming a more publicly and democratically distributed and employed process that nonetheless engages artefactual, material points of productive friction
- Design that is contextually grounded and stands more chance of being better-connected to local practices and identities
- The constitution of new socialities and spatialities that may cut across bureaucratic and administrative structures
- The potential for a radical redirection of citizenry away from the dominance of commercial interests to collective, societal concerns that might include processual and artefactual routes into engaging practically with such challenges as climate change, social justice and food security

This is the good news – the best-case scenario for design citizenship. But this isn't necessarily how it might be played out in reality. There are a number of ways by which the ambitions of innovative governance arrangements may not necessarily be reached. Following Swyngedouw (2005: 2002–3), these are threefold. First, governmental interests may privilege certain networks, over others. While it may appear to devolve out decision-making to groups outside its own bureaucracies, one has to consider which groups it is choosing to foreground over others. Second, tensions will occur between more emancipatory, transparent and participative approaches and systems of regulation, technocracy and power-based interests. How are these tensions to be resolved? Third, the tactic of promoting on 'ambassadors' to champion new forms of citizenship raises the question of representation. Who are they acting for other than themselves? What groups that were previously represented through other systems are now excluded in this new system?

Design is unavoidably implicated in these challenges. It may be mobilised as a tool through which these questions may be confronted (Latour and Weibel 2005), but it has to be understood as just one tool among others that are in circulation and use. A critical view on its limitations and affordances has to be pursued at the same time. Likewise, if we are to embrace design citizenship as an ontological mode, then how it is enacted, for whose interests and with what tools to hand are crucial issues. This problematisation may, in turn, arrest the flow of enthusiasm to a point of stasis where traditional hierarchies and routines are allowed to continue.

6.8 Conclusion

The coupling of design and the urban in discourses of policy and planning immediately rarefies and restricts how the former operates. It pulls it into questions of city boosterism and promotion, clustering models of economic development and creative industries, the role of design in the urban public realm and its involvement in wider processes of the use of culture to draw financial and human capital into cities. As such, it becomes an instrument of spatial planning wherein the potential of design to open other discursive fields, scales and forms of governance or everyday practices may be lost.

This may be regarded, however, as a historical moment, reflecting growth strategies in relatively stable economic times. The ground for this had already been through the promotion of New Public Management in the 1980s and 1990s followed by tentative movements to network governance from around 2000 onwards. New alliances and hybrid possibilities for connecting state administrations and citizens emerged. Following the financial crisis following 2007–2008, new spaces for design and governance to interact have been produced. Austerity measures forced the hand by placing increased emphasis on achieving end results of services over the bureaucracies of their delivery.

This has been met by a growth in design specialisms that are concerned with an entire systems approach rather than the fashioning of particular, isolated objects: this includes aligning user needs with organisational structures of delivery and paying attention to the orchestration of objects and people that constitute this. Again, this loosens the traditional ways by which designers work, offering more flexibility and greater breadth in their mode of intervention. This loosening is paralleled in modes of governance. Soft spaces disrupt administrative systems that neatly define territories; they work across and between these, and at the same time, new alliances and patterns of decision-making and implementation are founded.

The concept of design citizenship should therefore be read in the context of these developments. At the larger level, it signifies the active engagement of citizens in debate as to the material and visual qualities and meanings of the public realm, as in the case of Barcelona in the 1980s and 1990s. But it may also be possible to push beyond this to conceive of a form of design citizenship that is more closely embedded into everyday actions and decision-making. This is where design thinking is implicated into civic processes such as education curricula and organisation, social welfare and healthcare as well as economic development processes. Such is the case of Kolding from 2012.

These are not the only scales at which design citizenship is mobilised. This chapter has mostly discussed spatial planning and design citizenship in relation to larger scales such as cities and the broader territorial and administrative disruptions of soft spaces of governance. But design citizenship can also be enacted in more localised levels, consonant with Allmendinger and Haughton (2010) notion of bottom-up planning where 'official' boundaries and scales are not of concern. This is also where communitarian initiatives provide other scenarios for the coarticulation of practices and politics (Marres 2011, 2012). Community gardens, solar energy co-ops, self-build co-housing, neighbourhood time-banking schemes, reciprocal care systems, localised food networks, craft groups: these and many other collective activities provide bottom-up, situated platforms for design citizenship. While the municipally driven examples of Barcelona and Kolding functioned on a macro, citywide scale, these micro-scale examples attend to the 'dignity of the particular' as described by Kant and developed by Adorno (Stark 1998). But this is also where friction with instrumentalist top-down planning occurs. The spatial-material-discursive sedimentation of neoliberalism – its private-public urban design schemes aimed at city boosterism and maintaining property values, its shopping streets, corporately sponsored ice rinks and Ferris wheels, its riverside walks that figure a cacophony of chain restaurants and bars – occasionally comes into conflict with more communitarian practices. Soft spaces of governance and design citizenship have multiple politics and should therefore not be regarded as a single field.

Finally, it is important to reiterate that this shift away from spatial planning to design citizenship, via soft spaces of governance, engages a different temporality. The *front-lining* of public sector innovation through OBB or OBC along with 'results-driven design' certainly provides, conceptually, for more tightly focused spatial conceptions as it focuses more sharply on the interface of use. At the same time, these also become sites of experimentation and prototyping that are inherently unstable, contingent and forward-facing. Participation in this is part of the process of design citizenship – a continual state of becoming.

References

Allmendinger P, Haughton G (2009) Soft spaces, fuzzy boundaries, and metagovernance: the new spatial planning in the Thames Gateway. Environ Plan A 41(3):617–633

Allmendinger P, Haughton G (2010) Spatial planning, devolution, and new planning spaces. Environ Plan C Gov Policy 28(5):803–818

Anholt S (2002) Foreword. J Brand Manag 9(4–5):229–240

Bell M, Jayne M (2003) 'Design-led' urban regeneration: a critical perspective. Local Econ 18(2):121–134

Bennett R, Savani S (2003) The rebranding of city places: an international comparative investigation. Int Public Manag Rev 4(2):70–87

Berglund E (2013) Design as activism in Helsinki: notes from the world design capital 2012. Des Cult 5(2):195–214

Boddy M, Parkinson M (2004) City matters: competitiveness, cohesion and urban governance. Policy Press, University of Bristol, Bristol

Boland R, Collopy F (eds) (2004) Managing as designing. Stanford University Press, Stanford

Brandsen T, Boogers M, Tops P (2006) Soft governance, hard consequences: the ambiguous status of unofficial guidelines. Public Adm Rev 66(4):484–657
Brecknock R (2006) More than just a bridge: planning and designing culturally. Comedia, London
Buck N, Gordon I, Harding A, Turok I (eds) (2005) Changing cities: rethinking urban competitiveness, cohesion and governance. Palgrave, London
Busquet J (1993) Cobi al Descobert. Parisfal Ediciones, Barcelona
Calavita N, Ferrer A (2000) Behind Barcelona's success story: citizen movements and planners power. J Urban Hist 26(6):793–807
DCMS (Department for Culture, Media & Sport) (2014) Creative industries economic estimates, January 2014, statistical release. DCMS, London. https://www.gov.uk/government/statistics. Accessed 7 Mar 2015
Delgado M (2007) La ciudad mentirosa: fraude y miseria del 'modelo Barcelona'. Catarata, Barcelona
Deserti A, Rizzo F (2014) Design and the cultures of enterprises. Des Issues 30(1):36–56
Design Week (2009) An interview with LG head of design Luke miles. http://www.designweek.co.uk/issues/12-march-2009. Accessed 6 Mar 2015
Drew C (2015) Prototyping an online crime reporting service: a policy lab success story. https://openpolicy.blog.gov.uk/2015/02/03/prototyping-an-online-crime-reporting-service-a-policy-lab-success-story. Accessed 26 Feb 2015
Evans G (2002) Cultural planning: an urban renaissance? Routledge, London
Evans G (2003) Hard-branding the cultural city – from Prado to Prada. Int J Urban Reg Res 27(2):417–441
Evans G (2009) Creative cities, creative spaces and urban policy. Urban Stud 46(5–6):1003–1040
FIRJAN (2014) Mapeamento Da Indústria Criativa No Brasil. Rio de Janeiro: Sistema. FIRJAN, Rio de Janeiro, http://www.firjan.org.br/economiacriativa. Accessed 07 Mar 2015
Florida R (2002) The rise of the creative class: and how it's transforming work, leisure, community and everyday life. Basic Books, New York
Fortier A-M (2010) Proximity by design? Affective citizenship and the management of unease. Citizenship Stud 14(1):17–30
Gonzaga S (2012) From a design museum towards a European cultural place: the design milieu: strategies for European design culture in the globalization era. PhD thesis, Politecnico di Milano
Gordon I, McCann P (2000) Industrial clusters: complexes, agglomeration and/or social networks. Urban Stud 37(3):513–532
Hartley J (2005) Innovation in governance and public services: past and present. Public Money Manag 25(1):27–34
Hassan G, Mean M, Tims C (2007) The dreaming city: glasgow 2020 and the power of mass imagination. Demos, London
Haughton G, Allmendinger P, Counsell D, Vigar G (2010) The new spatial planning: territorial management with soft spaces and fuzzy boundaries. Taylor & Francis, London
Haughton G, Almendinger P, Oosterlyn S (2013) Spaces of neoliberal experimentation: soft spaces, postpolitics and neoliberal governmentality. Environ Plan A 45(1):217–234
Haughton G, Deas I, Hincks S (2014) Making an impact: when agglomeration boosterism meets antiplanning rhetoric. Environ Plan A 46(2):265–270
Jiménez A (2014) Introduction: the prototype: more than many and less than one. J Cult Econ 7(4):381–398
Julier G (2005) Urban designscapes and the production of aesthetic consent. Urban Stud 42(5–6):869–888
Julier G (2009) Designing the city. In: Julier G, Moor L (eds) Design and creativity: policy, management and practice. Berg, Oxford, pp 40–56
Julier G, Leerberg M (2014) Kolding – we design for life: embedding a new design culture into urban regeneration. Finn J Urban Stud 52(2):29–56
Kauko T (2010) Sustainable urban property development and neighbourhood dynamics. Open Urban Stud J 3:103–111

Keane M (2013) Creative industries in China. Polity Press, Cambridge
Kimbell L (2011) Rethinking design thinking: part 1. Des Cult 3(3):285–306
Kimbell L (2012) Rethinking design thinking: part 2. Des Cult 4(2):129–148
KPMG (2011) Making the transition: outcome-based budgeting: a six nation study. http://www.kpmg-institutes.com/content/dam/kpmg/governmentinstitute. Accessed 27 Feb 2015
Laclau E, Mouffe C (2001) Hegemony and socialist strategy: towards a radical democratic politics, 2nd edn. Verso, London
Lacroix M-J (ed) (2005) New design cities. Les éditions Infopresse, Montreal
Latour B, Weibel P (eds) (2005) Making things public: atmospheres of democracy. MIT Press, Cambridge, MA
Law J (2013) Do outcomes based approaches to service delivery work? Local authority outcome agreements. University of South Wales, Centre for Advanced Studies in Public Policy, Wales. http://caspp.southwales.ac.uk/media/files/documents/2013-06-24/publication-local-authority-outcome.pdf. Accessed 27 Feb 2015
Manzini E, Jégou F (2005) Sustainable everyday: scenarios of urban life. Edizioni Ambiente, Milano
Marres N (2011) The costs of public involvement: everyday devices of carbon accounting and the materialization of participation. Econ Soc 40(4):510–533
Marres N (2012) Material participation: technology, the environment and everyday publics. Palgrave Macmillan, Basingstoke
McCarthy J (2007) Partnership, collaborative planning and urban regeneration. Ashgate, Aldershot
Michlewski K (2008) Uncovering design attitude: inside the culture of designers. Organ Stud 29(3):373–392
Narotzky V (2007) La Barcelona de Diseño. Santa & Cole, Barcelona
Olesen K (2012) Soft spaces as vehicles for neoliberal transformations of strategic spatial planning? Environ Plan C Gov Policy 30(5):910–923
Papadopoulos N, Heslop L (2002) Country equity and country branding: problems and prospects. J Brand Manag 9(4–5):294–315
Peck J (2005) Struggling with the creative class. Int J Urban Reg Res 29(4):740–770
Press M, Cooper R (2003) The design experience: the role of design and designers in the twenty-first century. Ashgate, Aldershot
Reimer S, Pinch S, Sunley P (2008) Design spaces: agglomeration and creativity in British design agencies. Geogr Ann Ser B Hum Geogr 90(2):151–172
Saco R, Gonsalves A (2008) Service design: an appraisal. Des Manag Rev 19(1):10–19
Schmitter P (2002) Participation in governance arrangements: is there any reason to expect it will achieve 'sustainable and innovative policies in a multi-level context'? In: Grote J, Gbikpi B (eds) Participatory governance: political and societal implications. Leske and Budrich, Opladen, pp 51–69
Seaton A (1999) Book towns as tourism developments in peripheral areas. Int J Tour Res 1(4):389–399
Stark T (1998) The dignity of the particular: Adorno on Kant's aesthetics. Philos Soc Crit 24(2–3):61–83
Stevenson D (2004) 'Civic gold' rush. Int J Cult Policy 10(1):119–131
Stickdorn M, Schneider J (eds) (2010) This is service design thinking. BIS Publishers, Amsterdam
Storper M (1997) The regional world: territorial development in a global economy. Guildford Press, New York
Sum N-L (2010) The cultural political economy of transnational knowledge brands: Porterian competitiveness discourse and its recontextualization to Hong Kong/Pearl River Delta. J Lang Polit 9(4):546–573
Sung W, Song M, Park J, Chung K (2007) Changing roles of design promotion organizations in the global context and a new theoretical model for a design promotion system. International Association of Societies of Design Research Conference Hong Kong. http://www.sd.polyu.edu.hk/iasdr/proceeding/papers. Accessed 6 Mar 2015

Swyngedouw E (2005) Governance innovation and the citizen: the Janus face of governance-beyond-the-state. Urban Stud 42(11):1991–2006

Syrett S, North D (2008) Renewing neighbourhoods: work, enterprise and governance. Policy Press, Bristol

Temporal P (2006) Asia's star brands. Wiley, Chichester

Turner F (2006) From counterculture to cyberculture: stewart brand, the whole earth network, and the rise of digital utopianism. University of Chicago Press, Chicago

UNCTAD (2010) Creative economy report 2010. United Nations, Geneva. http://unctad.org/en/pages/PublicationArchive.aspx?publicationid=946. Accessed 7 Mar 2015

UNESCO (2015) Creative cities network: why cities? http://www.unesco.org/new/en/culture/themes/creativity/creative-cities-network/why-cities. Accessed 27 Feb 2015

Urban Task Force (2005) Towards a strong urban renaissance. http://www.urbantaskforce.org/UTF_final_report.pdf. Accessed 6 Jan 2014

Weber C (2010) Introduction: design and citizenship. Citizenship Stud 14(1):1–16

Zhang H (2008) Shenzhen: frontier city. In: Zhang H, Parker L (eds) China design now. V&A Publishing, London, pp 42–52

Part III
Innovating Governance and Economic Models

Chapter 7
The Frugal Government Concept Revisited

Francesco Molinari

Abstract Revisiting the frugal government concept is the aim of this chapter, which first provides a working definition and then discusses the renewed importance of it for a government modernization agenda. Frugal basically means parsimonious (rather than simply efficient or effective) in the use of resources, inclusive and participatory with both civil society and individual citizens, and open to public service restructuring with and by the active contribution of beneficiary people and communities. The concept is further analyzed in its implications for service renewal in relation with the dynamics of personal engagement, collective learning, and behavioral change.

Keywords Public management • Inclusive government • Service cocreation

7.1 Motivation

> A wise and frugal Government, which shall restrain men from injuring one another, shall leave them otherwise free to regulate their own pursuits of industry and improvement, and shall not take from the mouth of labor the bread it has earned. This is the sum of good government, and this is necessary to close the circle of our felicities. (Jefferson 1801)

This quote from Thomas Jefferson's inaugural address as third president of the United States, delivered at the Capitol Building, in Washington, DC, on Wednesday, March 4, 1801, is deemed to be the first (and is certainly the most popular) summary of what frugality in government should mean, a principle, or rather a set of *essential principles*, which ought to shape any public administration, as proposed by Jefferson, who felt the need to share this commitment with his fellow citizens, being about to enter *on the exercise of duties which comprehend everything dear and valuable to you.*

Since then, the ideal type of frugality has rarely been used in combination with a shared (or majoritarian) concept of government performance, either in theoretical or pragmatic reflections. About two centuries later, pushed by the "'New Public

F. Molinari (✉)
Department of Architecture and Urban Studies, Politecnico di Milano, Milan, Italy
e-mail: mail@francescomolinari.it; francesco.molinari@polimi.it

G. Concilio, F. Rizzo (eds.), *Human Smart Cities*, Urban and Landscape Perspectives 19,
DOI 10.1007/978-3-319-33024-2_7

Management" (NPM) doctrinal revolution of the 1980s, frugality reappeared in a list of seven key dimensions of government transformation, listed in two seminal papers by Christopher Hood (1991, 1995), implicitly suggesting that stasis had been the rule thus far. In that context, a more parsimonious use of public resources (or in Jefferson's words, "more bread left in the mouth of labor") was seen as part of a maturing process toward budget discipline, enabling to deliver services that were stable in terms of volume and, therefore, not impacting on user appreciation and satisfaction.

In such a vision, then, frugality could somehow be considered as synonymous of process efficiency or the capacity of the public administration's "factory" to produce a stable level and quality of services (namely, its industrial outputs) with a reduced or diminishing amount of resources. Associating service delivery to public action may seem fairly obvious today, particularly as far as the local government is concerned, but the emphasis on cost-effective fulfillment of goals was quite novel at the time and a distinctive feature of NPM compared with its historical predecessor, the bureaucratic practice, and functional theory of administration (see Dunleavy and Hood 1994).

Twenty more years have passed – after the size and composition of public finance were heavily hit by the financial crisis, with very few exceptions worldwide – efficiency-focused government transformation efforts have been reconsidered against the very real necessity of downsizing public administration to avoid the risk of closing it altogether. In this new scenario, we posit that a renovated attention to frugality may be an interesting alternative – for various reasons, which will be explained later in this chapter – to the sheer reduction in the number and/or adjustment in the quality of those public services that are no longer sustainable with their former, old-style production and delivery process. Like in Jefferson's talk, a new perimeter of public action could be designed, which leaves men and women "free to regulate," not only their respective "pursuits of industry and improvement" but also the extent of their own engagement in individual and joint endeavors, ultimately aiming to innovate public service in response to precise citizen needs.

Interestingly enough, this new push toward frugality in government (and more generally speaking, product/service engineering, in the private sector as well) comes from developing countries. India's Tata Nano, known as the cheapest car in the world, for sale at the price of a motorbike (about $2,500), resulted from numerous radical innovations in design, engineering, and manufacturing, guided by three key requirements: value for money, compliance with regulations, and acceptable performance standards (Tata Motors 2015). GE's hand-carried healthcare ultrasound system, Logiq Book, developed for use in China's rural areas, has significantly fewer features than traditional ultrasound devices. However, it costs almost 80 % less than competitor products and is much smaller and lighter, and its portability means that rural patients do not have to travel to faraway cities to receive diagnosis and treatment (Immelt et al. 2009). In Kenya, a joint venture of Vodafone UK and Kenyan Safaricom invented a mobile microfinance service called M-Pesa for people who do not have a mobile bank account. The service uses existing mobile phone technology (SMS) and infrastructure, allowing registered users to load money on their device, which can then be sent via SMS. The recipient of the text message can pick up the cash at his/her nearest vendor. Over 50 % of the adult population in Kenya use the service to send Pesa (standing for money in Swahili) to far-flung relatives and to pay for shopping and utility bills (Graham 2010).

While the former examples of (radical as well as) frugal innovation come from a top-down initiative of private enterprises, the UK Serco Institute's study on over 40 social businesses from India allowed to conclude that *citizens do not care who delivers the service, the government or for-profit or not-for-profit organizations, so long as it meets their requirements in terms of outcomes and quality* (Singh et al. 2012). There is therefore plenty of room for public service redesign and reengineering to capture a higher number of beneficiaries with new or improved solutions, which seriously meet user needs and requirements while cutting down service delivery costs considerably. In India, where citizens rarely receive public services free of charge and many have to survive without even an equitable access to them, the impetus for local people to embrace frugal innovation came from this sense of (and actual) exclusion. In the Western world, where the delivery of high-quality and universal services is a long-standing tradition, which has only just become unaffordable, the new challenge is how to exploit innovation from the bottom-up to continue ensuring everybody has the right to access services in a fair, transparent, and uniform way.

The structure of this chapter is as follows. The next section provides a working definition of frugal government, aligned with the ideas presented above. The definition is then further analyzed with its implications for the logic model of government, which is the topic of the third section. A discussion follows on the rationale and implications of restructuring public service delivery according to frugal principles. In particular, the connections between participation, learning, and behavioral change are presented. Finally, a discussion on "true participation" as a recurrent game is presented. Some conclusions are drawn in the last section of the chapter.

7.2 What We Mean by Frugal Government

First and foremost, it is necessary to emphasize a slight misunderstanding that has occurred among both researchers and practitioners regarding the use of three words in relation to government: lean, efficient, and frugal.

Usually "lean government" is used to refer to the application of "lean production" principles and methods to the delivery of public services. The original lean production expression was coined by a former quality engineer at the Toyota-GM joint venture NUMMI in California, in a 1988 article based on his master's thesis at the MIT Sloan School of Management (Krafcik 1988). It may actually refer to two distinct approaches to quality enhancement in manufacturing, one focused on the elimination of waste, the other on improving the "smoothness" of adopted workflow(s). Likewise, lean government approaches point to identifying, and then removing, the administrative bottlenecks and useless superstructures that do not add value for the citizens, but simply keep the organizational teams engaged in noncore activities and negatively affect the perceived quality, speed, and transparency of public action. At face value, it may seem that lean government practices are great contributors to frugality. In fact, international evidence displays the following paradox: the countries where the majority of public opinion perceives a better quality of government are also those providing stronger support to increasing public expenditure and therefore taxation (Svallfors 2013).

With regard to efficient government, beyond its formal definition (once again borrowed from the manufacturing practice and presented in the previous section of this chapter), it can be added here that three distinct measures of efficiency have been introduced, historically, with the aim of documenting government performance in a (kind of) objective way:

- The ratio of public expenditure over GDP, considered as a sort of opportunity cost, against the alternative hypothesis of letting the market, or civil society, be self-organized in fulfilling the goals of public action without recurring to distortive taxation
- The same ratio, but over the number of served beneficiaries, which can be seen as a more refined metric of spending efficiency within a certain service or function, for instance, in the domain of healthcare or education
- A combination of efficiency and effectiveness indicators, such as the ones presented in a study for the European Central Bank by Afonso et al. (2003) and in another, quite similar, study conducted for the Inter-American Development Bank by Afonso et al. (2013)

Overall, international evidence is quite consistent in showing that no particular progress has been achieved under any of the above measures of efficiency by governments around the world. In fact, public expenditure has been steadily growing as a GDP ratio across the years, and, with very few praiseworthy exceptions, the cost per served beneficiary has also been growing over time in all the major functions of public administration. Therefore, it is possible to conclude that no contribution to frugality has been coming from the strive toward efficiency in government, if ever it materialized in real practice.

A qualification to this conclusion is provided by the third measurement stream mentioned above, which univocally points to the size of government as inversely correlated with its efficiency. More specifically, it has been found that the group of countries with public spending below 40 % of GDP reported better public sector performance than the one above 50 %, while the medium-sized national governments (spending between 40 % and 50 % of GDP) stayed in between. In other words, this quantitative benchmarking exercise seems to indicate that the capacity of being efficient in a government is positively correlated with its attitude to being frugal – i.e., parsimonious in global expenditure. This can be graphically represented as in Fig. 7.1.

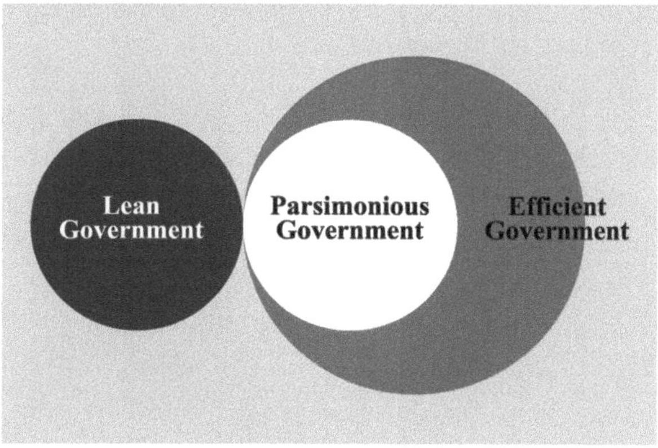

Fig. 7.1 Lean, efficient, and parsimonious government concepts

Putting together all the aspects above, we would like to propose the following definition: wise and frugal is a government that at the same time:

- Is parsimonious in the use of resources, not only financial – if and when they exist – but also including the creativity, imagination, and collaboration of people, from both inside and outside the public sector
- Aims to minimize the cost and maximize the benefit of public service delivery per user – therefore, it adopts a value-centered perspective, where value is defined in differential terms between benefit and cost, and these are both expressed in nonmonetary (subjective) and monetary (objective) terms
- Adopts the diligence of a good family man in finding, measuring, and distributing available (scarce) resources in the direction of societal transformation – including to empower, rather than prevent, the activism and behavioral change of all the members of its own, extended "governance system," i.e., citizens, businesses, third sector operators, academics, media, and other stakeholders (Fig. 7.2)

Appropriate examples, such as the ones listed below, can be used to clarify the above statements:

- A more parsimonious use of human resources avoids the risk of "consultant lock-in": whenever a public agency commits to some external expertise to perform certain tasks, it should set making its internal staff increasingly autonomous (both conceptually and operationally) as an additional goal. This has little to do with efficiency in spending, unless it is seen as a way to reduce the recourse to such external expertise across time.
- Public value is normally used as an abstraction (and sometimes as a proxy) of government performance. However, this is not a monolithic concept, but a multifaceted one – considering both the variety of nuances associated with its significance (from wealth to health, from public safety to quality of life, from social justice to equal opportunities) and the variety of its meanings as expressed by the different stakeholders materially affected by public action (Harrison et al. 2011).

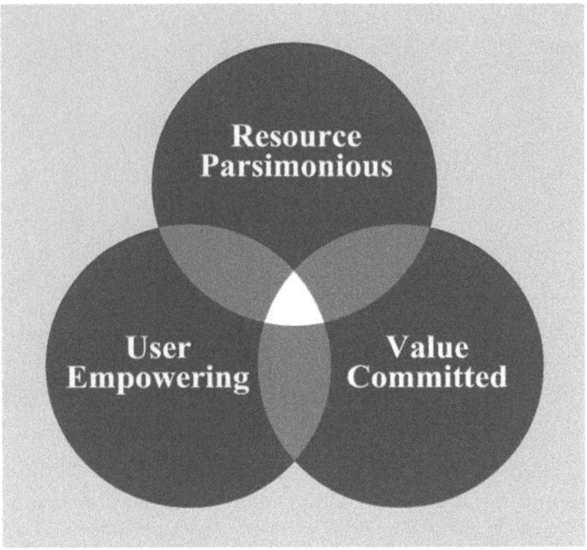

Fig. 7.2 Our definition of "wise and frugal" government

- The concept of public value is also dynamically changing, on both sides of the government-citizen relationship. On the one side, population dynamics and demographic trends have created new dilemmas and potential tensions between the interests of younger and older generations, for instance, within the pension and healthcare systems of many Western countries. On the other side, as shown by the study of Singh et al. (2012) quoted in the introduction, in emerging countries like India, over the past 60 years, social enterprises have stepped in to address the challenges that the government was not dealing with. This has promoted and supported radical perspectives pushing for a limited intervention or involvement of the state in public affairs.
- To be user empowering does not necessarily mean, however, that frugal governments should abstain from directly contributing to the organization of service production. In fact, what changes here is basically the role played by leadership, switching from bridging social capital in the community to creating tighter relationships between stakeholders via a more direct engagement in doing things (Beer and Clower 2014).
- As Szkuta et al. (2012) put it, collaborative production of public services with and by the users allows perceived quality to rise with the number of people involved in it, rather than not varying (as occurs with electronic services) or diminishing (with traditional ones). This is a fundamental add-on to the value proposition of any such service, which does not depend on the government's organizational capacity, but only on the extent of involvement of the private sector.
- A more inclusive, dialogic, and collaborative approach with the end users of a certain service also generates new ideas, concepts, and recommendations that help improve perceived quality while keeping cost and other industrial parameters under control. According to the theories of crowdsourcing (Fleming

2004; Surowiecki 2004), diversity of opinions, independence of judgments, use of local knowledge, and fair aggregation of preferences are the indispensable ingredients to achieve user-driven innovation.

7.3 The Logic Model of Frugal Government

From the way we have presented it, our definition of frugal (and wise) government is inextricably related to societal transformation. This marks a clear distinction with other, merely incremental, theories of public sector innovation, such as the Transformational Government Framework (TGF), which promotes IT adoption to improve the delivery of public services (CS Transform 2010). Additionally, it challenges the policy maker on how to create the best recipe with the available ingredients.

While some early examples of managerial "cookbooks" have started to appear (see Molinari et al. 2013; Marsh et al. 2014; Eskelinen et al. 2015), none of them explicitly links their recipes and recommendations to the implementation of the frugal government concept.

To make a step forward in this direction, we propose adopting the program logic model introduced by the WK Kellogg Foundation (1998, 2004) in support to its evaluation exercises (see Fig. 7.3). The five building blocks of the program logic's visual representation are quite familiar to policy makers, namely, consisting of:

- *Resources or inputs*: such as money, staff, and equipment, which are required to operate a program
- *Activities*: the program implementation work made possible by an appropriate use of the resources

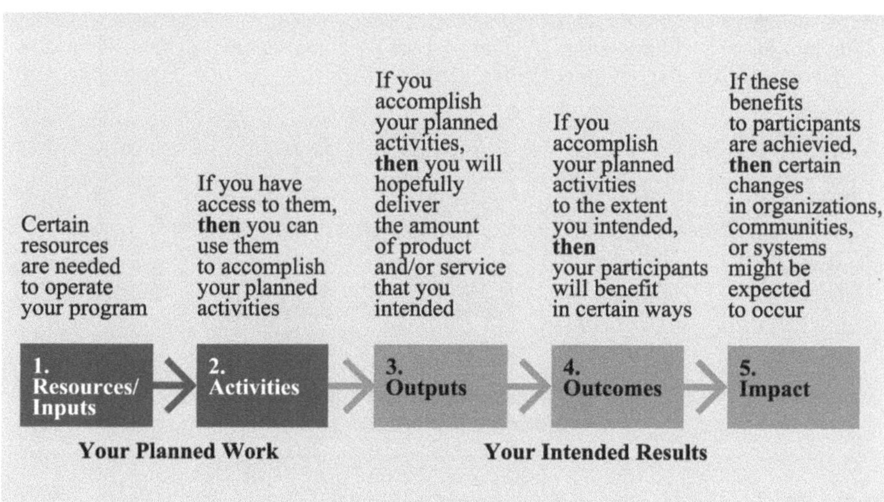

Fig. 7.3 The program logic model by the Kellogg Foundation

- *Outputs*: the "industrial" results delivered by the program at its end, in terms of, e.g., products and/or services
- *Outcomes*: the benefits brought in by the outputs to, e.g., the participants in the program or served beneficiaries
- *Impact*: the expected/achieved degree of change in the targeted individuals, organizations, communities, or systems

Ideally, for each stage of this stepwise conceptual workflow, a number of key performance indicators can be gathered, and related analyses/implications/lessons can be drawn, such as:

- Relevant measures for the use of human, financial, and organizational resources at stage 1
- Activity-level indicators for each planned task to be carried forward during stage 2
- Quantitative delivery metrics for each product/service of interest at stage 3 of the logic model
- Satisfaction indexes and other quality assessment parameters/exercises at stage 4
- Impact and change proxies (due to lack of actual observations, concurred driving factors, time lags, and fuzzy propagation pathways) at stage 5

For the purposes of this chapter, we will now use the above representation to outline the main differences between the traditional and the emerging, frugal, "government's business" logic, as shown in Fig. 7.4.

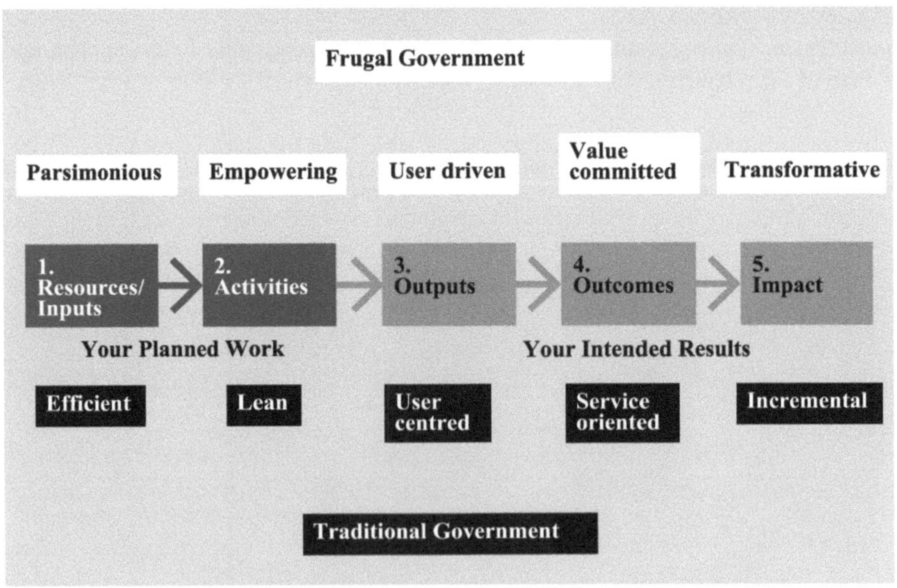

Fig. 7.4 The logic model of traditional and frugal government

More specifically, we note that:

- At stage 1, the distinction between efficiency and parsimony has been already introduced and commented above. Not all governments that can be said to be efficient are also parsimonious, while the opposite is probably true. This is to be reflected in the overall setup and organization of public administration and/or service delivery activities, as implied below.
- Likewise, striving toward more inclusion and participation of citizens and stakeholders, on top of the quality enhancement of public action, which characterizes stage 2, has already been discussed. Other things being equal, perceived service quality by its beneficiaries is higher if/when there is room for codesign or co-delivery, compared with any other possible attitude to lean production. However, the essence of this new managerial concept is that users or beneficiaries of public services should not only be included in the process but also empowered to it (Denhardt and Denhardt 2000), which will lead to the positive outputs described below.
- At stage 3, the step leading from conventional modernization programs to frugality in government can be better appreciated if one looks at the familiar distinction between user-centered and user-driven innovations, which lies at the core of interaction design (Cooper 2004) and has been pushed to the limit by the living lab approach (Pallot et al. 2010; Dell'Era and Landoni 2014). In the area of public service – as of software development – the point is not to generate outputs that conform to an idealistic (or idealized) vision of the customer, but to elicit his/her direct engagement in the design and delivery process, on a peer basis with the service provider, which will add an original touch to the obtained results.
- In stage 4, the benefits brought to participants in service cocreation (a term covering both codesign and co-delivery) have been shown to extend far beyond the quality enhancement or improvement/innovation of a single public service instance. In particular, peer collaboration between public and private actors, the latter also including individual citizens, leads to enhanced societal value (Hui and Hayllar 2010), by leveraging on a number of concurrent factors, such as cultural convergence between involved parties, better process knowledge and "ownership" among service beneficiaries, and an increased sense of belonging or identification.
- At the final stage 5, the logic model of frugal government displays its very essence, which is to be deeply transformative rather than marginally incremental with respect to the status quo ante. The reason is straightforward. What is ultimately aimed at (or required) is the capacity to "do more with less" and "no more of the same" (Accenture 2008) under the double pressure of decreasing financial resources and the growing complexity of the socioeconomic and environmental problems at hand.

The vision of collaborative public sector innovation has been around for quite a while now (see Agranoff and McGuire 2004; Bommert 2010). However, it is only with the coming of information and communication technologies (ICTs) that governments around the world have been able to carry out both top-down and bottom-up experiments aiming at a much more extended user integration than conventional "off-line" approaches. In doing so, collaboration with beneficiaries has started to be perceived as potentially far more effective in transforming the public sector's "old-style business."

For instance, the European Commission's eGovernment Action Plan 2011–2015 (European Commission 2010) brought about the concept of open, flexible, and collaborative service delivery, putting emphasis on a number of good practice examples and viable approaches, including across the borders, which the EU member states could take inspiration from (European Commission 2010). In June 2013, the public service unit of DG CNECT drafted a vision document with the aim of further highlighting the way ICT can help European governments to empower ordinary citizens and entrepreneurs, in making their voices be heard and allowing them to share some of the conventional public sector tasks, regardless of distance and at any moment in time (European Commission 2013).

Public/private collaboration in the design, production, and validation of public services is only one aspect of the proposed open government paradigm (see Fig. 7.5), the others being transparency and participation in collective decision-making processes. The same three principles lie at the heart of the Obama Administration's Open Government Directive (Orszag 2009).

Were it not for the digitalization aspect, the essential contribution of end users to the productivity, efficiency, and operational quality of service-providing organizations would be a known fact in the marketing and management literature over the past 50 years or so (Fuchs 1968). For instance, the diagram in Fig. 7.6 (borrowed from Büttgen and Ates 2009) maps a huge number of conventional, i.e.,

Fig. 7.5 The open government paradigm

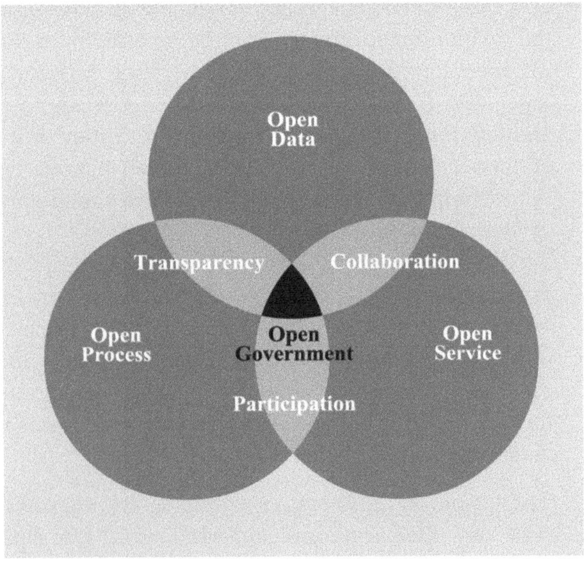

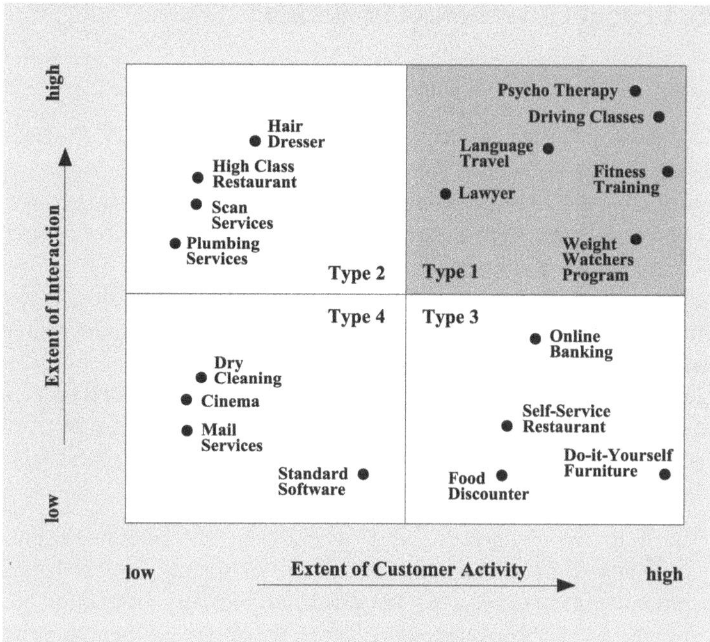

Fig. 7.6 Taxonomy of services according to customer participation

nonelectronic, services in relation to the extent of customer activity during delivery and to the degree of one-to-one, personal interaction between user and provider. In the top right quadrant, several "type 1" services are mentioned, in which the client or customer contributes significantly to the production and delivery process and its outcomes.

However, from this relatively simple picture, many interesting variants are left out, including the impact of:

- Multiple users and/or service providers, rather than a single one, involved in the interaction at production and delivery level.
- Individual end users or communities playing the role of service providers themselves. This is often referred to as the "prosumer" (producer + consumer) case.
- A more refined distinction between services that are natively interactive, or collaborative, and those that become so after a dedicated reengineering effort by the provider.
- A deeper consideration of the rationale, or motivation, for a government agency to integrate users in the service delivery process.
- An analysis of the contextual conditions making it more likely, or viable, for a public service provider to become frugal – and wise – over time.

These aspects are dealt with globally in the following section.

7.4 The Frugal Government in Action

Briefly stated, integrating citizens/customers in the joint delivery of services between the public and the private sector is the main innovation of the frugal government concept. This generates at least three positive effects:

1. New, more parsimonious approaches to service delivery, exerting lower pressure on public finance and therefore avoiding the risks of reduction in scope, if not complete dismissal, of the service because of budget crises
2. Increased level of engagement, and therefore satisfaction, among the citizens (individuals and/or groups) collaborating with the service production and delivery process
3. The establishment of a new wave of cocreated and codeveloped services, which fulfill existing or new requirements emerging from the citizens, in a smarter and more inclusive way. Following a recent literature stream, we call these "Human Smart" government services (Concilio et al. 2015).

While the first two goals could also be grasped with alternative public management strategies (outsourcing and privatization in one case, subsidiarity and lean government in the other), only the third one is really connatural to the public/private/people partnerships that characterize frugal government, and it therefore requires extra motivation and justification.

The rationale of implementing a human perspective to public service development lies primarily in the growingly "wicked" nature of the societal problems that these services are supposed to tackle. The use of this term started at the end of the 1960s in the context of social policy and design science (Churchman 1967; Rittel and Webber 1973). It means "resilient to resolution" – in contrast to "tame," easy-to-solve problems – rather than "evil" or "dreadful" as the word suggests.

Wicked problems cannot be tackled by a purely scientific-rational approach (which would define, analyze, and solve them in sequential steps) for three major reasons:

- It is impossible to start with a clear and univocal problem definition, which depends on the solution framework and the different perspectives of the involved stakeholders.
- The constraints the problem is subject to and the resources needed to solve it change over time.
- There is no "right" or "optimal" solution, which actually depends on the proposed problem framework, and the problem can never be definitively solved.

Classic examples of wicked problems include economic, environmental, and political issues. A problem whose solution requires a great number of people to change their mind-sets and behaviors is likely to be a wicked problem (Allan 2011). Indeed, many wicked problems local governments worldwide are facing come from areas of interest (and often legal competence) such as waste recycling, water and energy saving, collective mobility, public safety, health, and social care.

The ideal type of services required to tackle these complex problems conforms to (at least) three main characteristics:

1. They generate some degree of behavioral change in the people involved in them.
2. They activate mass multiplication or viral diffusion effects, meaning that their outcome is more than simply the sum of individual behavioral changes.
3. They are sustainable – not only financially, which would also be good per se, but also from the societal and institutional points of view.

Not surprisingly, most of these application domains overlap with the fields of intervention for Smart City governments (Neirotti et al. 2014). This seems to imply that by making wide recourse to the technical "smartness" of sensors, meters, and ICT infrastructure, it would become easier to cope with those wicked problems in a satisfactory manner. By contrast, the (self-defined) Human Smart City approach recommends balancing technology injections and deployments with "softer" features such as social dialogue, collective vision building, people empowerment, and government-to-citizen interaction in physical (as well as virtual) community settings. The landing point will then be the design, development, and validation of new urban services through the application of user-driven and participatory methods.

The EU-funded project PERIPHÈRIA (www.peripheria.eu) carried out several experimentations focused on the creation of these Human Smart City services, with the participation of several stakeholders, including policy makers, city officials, technology providers, and citizens, each with different roles and capabilities. The key lessons learned from the project pilots (localized in six European cities, Athens, Bremen, Genoa, Malmoe, Milan, and Palmela) include the identification of four main drivers of behavioral change, namely, the following:

(a) *Personal interaction with new service platform(s)*. First and foremost, individuals and communities experienced the novelty of a service, which could contain elements (both material and immaterial) that made it substantially different compared to its alternatives. People were felt mocked, attracted, shocked, amused, entertained or even affronted, challenged, and disgusted by the new service experience. This had observable effects on subsequent behavior.

(b) *Individual and collective involvement in service codesign*. Citizens and stakeholders were strongly engaged in the service innovation process, through a user-driven open innovation ecosystem that unlocked their real needs, preferences, and aspirations and proposed to co-assemble solutions that looked very suitable for them. This infrastructure triggered participation in the cocreation of services, but also seemed to ignite a diffused, coherent, and sustained form of compliance in behavioral terms, for all those who had taken part in service codesign.

(c) *Individual and collective involvement in service coproduction*. In many cases, the service cocreated with the participatory methods and tools offered by PERIPHÈRIA was also a coproduced one. Namely, it required the active engagement of citizens and stakeholders – both as individuals and communities or social groups – to become effective and to be fully available to its expected beneficiaries. This had the power of reinforcing the compliance effect mentioned above, which was observable on those who took part in service coproduction.

(d) *Individual and collective involvement in service validation/evaluation*. All those services were born as experimental, as pilot platforms of some kind, requiring a period of time to be assessed – configured, installed, and populated with active end users – before reaching a stage of full maturity, thus allowing permanent deployment. The validation and verification – not to speak of evaluation – of those prototype platforms was an essential component of the participatory vision of PERIPHÈRIA. There were signals that after involving citizens and stakeholders in these activities, a permanent impact could be detected on individual and collective behaviors, again in terms of fuller compliance with the project's scope and the purposes that had driven its experimentations.

The missing link between collaboration (or participation) in service development and the resulting behavioral change has been identified in the process of *learning*.

Albert Bandura (1977), in his pathbreaking essay, stated that personal behavior is learned from the social environment through a process of observation of, and identification in, other "exemplary" human beings. More recently, Siemens (2005) proposed connectivism as an adjusted social learning theory integrating principles from chaos, network, complexity, and self-organization theories. The main sources of learning (defined therein as "actionable knowledge") reside outside us (e.g., within an organizational culture or database). Their effectiveness depends on our ability to retrieve and connect specialized information sets, including other people's advice, which do not belong to the learner's initial profile and range of contacts, but become relevant in the context of an informed decision-making process. This process is started and develops here and now, following sudden changes in the external environment, which impose our reaction. In this vision, the connections a person holds are more crucial to ensure learning than the knowledge assets one currently possesses.

Based on evidence from the PERIPHÈRIA project pilots, Concilio and Molinari (2014) argue that whenever new solution prototypes for "wicked" urban problems are cocreated with a strong contribution from citizens, the people involved in that process change their previous attitudes toward the common good. This is due to the production of relational capital, by the joint reflections and discussions on possible initiatives having a social nature and value for the community. As a result, people increase their civic awareness and become more effectively and permanently influenced in the direction of adapting/aligning their current and future behaviors to the emergent, shared vision of society.

Such virtuous learning does not only occur, however, within the boundaries of a well-identified group or organization. It is rather diffused among all the participants in the distribution of experimental tasks that shape the collective, place-based, action space (called "shared action arena" by Döös and Wilhelmson 2011).

How this takes place is the result of complex and concurrent dynamics of several distinct elements in an urban ecosystem. Indeed, the resulting solution takes on a socio-spatial nature, with a more distinct pattern than the outputs of individual or group creativity. This reinforces the perception that participation works better in smaller than bigger communities, while its impact at macro level is usually negligible, if any at all (Mansuri and Vijayendra 2013).

7.5 The Importance of "True Participation"

Service innovation strategies following the frugal and wise principles usually aim to engage all community stakeholders in order to find the best possible solution to "wicked" problems. Typically, this participatory process is not really scalable, or there is no easy way for it to reach higher aims (and figures of people involved in grassroots sessions) unless it is implemented at local level.

But what exactly do we mean by "participation"? The essence of it, quite often minimized or forgotten, is that a significant portion of civil society must be genuinely involved in shared decision-making with the representatives of public administration. These are the two necessary conditions for relational capital to be generated and, therefore, behavioral change induced. What actually happens is that either a small subset of civil society is mobilized or the kind of involvement has little to do with real co-decision-making (Skidmore et al. 2006). Loosening either condition overrules the transformational impact of user integration and therefore nullifies the benefits of the principal frugal government strategy.

Our proposal to solve this impasse starts from depicting participation as a "recurrent" game, engaging the local government (LG) and civil society (CS). Let us start by saying that the LG moves first. There would be two options at hand: seriously involving citizens in future decisions through an open and transparent process (Option 1) or acting in a ceremonial way, e.g., only promising and then not keeping to initial commitments (Option 2). In turn, CS could decide to attend the participatory process in an engaged and loyal way (Option 1), irrespective of what the LG does, or allow a very limited degree of engagement in operations, whatever they are (Option 2).

In the diagram in Fig. 7.7, we define the pairs (RLG_{ij}, RCS_{ij}) (i, j, = 1,2) as the rewards gained by each player in relation to the $2^2 = 4$ moves available.

In the case of CS, one could easily prove that $U(RCS_{21})<U(RCS_{22})$ for any cardinal and monotonic utility function and that $U(RCS_{11})>U(RCS_{12})$. Looking at the case of LG, it is also rather straightforward to show that $U(RLG_{21})<U(RLG_{22})$ and that $U(RLG_{11})>U(RLG_{12})$. Therefore, if the LG (who has moved first) plays fair, the fair play option is also preferable for CS. If on the other hand the LG decides to cheat, then the cheating option is the one preferred by CS. This, in our opinion, explains most of the failures of "top-down" participatory efforts that occurred in the past.

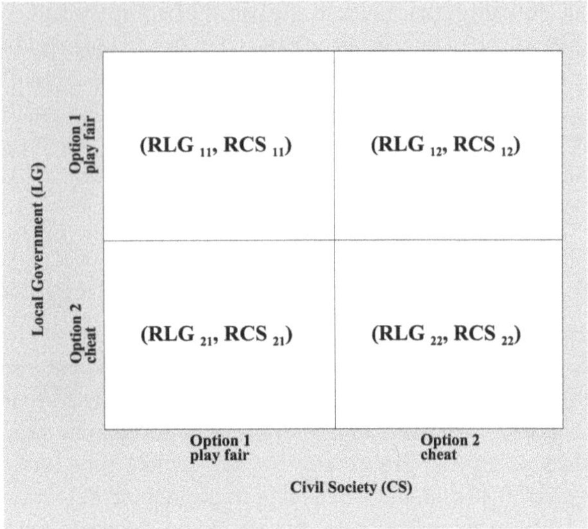

Fig. 7.7 The rewards of participation

In fact, how can the behavior of the LG be evaluated by CS before deciding how to react? Usually CS has no clue except the memory of past interactions with the LG. If it was cheating in the past, it could well be the case this time as well. Otherwise, the positive expectation would be reinforced. This is why we normally experience that, other things being equal, two consecutive rounds of a same game when the LG moves first could not finish with the two players acting in two different ways.

Now suppose that CS moves first. From the LG side, one can easily prove that $U(RLG_{11})<U(RLG_{21})$ and that $U(RLG_{12})<U(RLG_{22})$. In other words, irrespective of what CS is aiming to achieve, the cheating option is uniformly superior in the eyes of LG. From the perspective of CS, we should have $U(RCS_{11})>U(RCS_{21})$, while on the other hand $U(RCS_{12})<U(RCS_{22})$. In other words, the CS is running the risk while moving first, that the LG will cheat while CS plays fair, so that the result will be inconclusive.[1] Put differently, there is little chance for bottom-up participation to be taken seriously by an incumbent government, unless its set of priorities were radically changed – which is exactly what we mean when speaking about "introducing a culture of participation" in public administration.

Now let us change our previous example a little. This time, participation is no longer meant to "tie the government's hands" in assuming a specific decision that was jointly co-determined with the citizens or stakeholders. Now participation is somehow meant to "tie the citizens' hands" in forcing them to change their behavior in relation to a given issue or crucial standpoint, where the government's action alone could not be effective or decisive to the required extent. In this new example, the LG (contrary to the previous case) has little incentive to cheat, because it has

[1] We can figure that the case where the CS plays first and cheats would be rather unrealistic.

already experienced the necessity of a broader involvement of CS. So whenever the LG moves first, it would appear unrealistic that it decides to cheat as its primary option. Let us then look at what happens to the utility function of the CS. Would it still be the case that $U(RCS_{11})>U(RCS_{12})$, while also being that $U(RCS_{11})>U(RCS_{12})$, so that the "play fair" option would be the preferred one? That outcome is going to be somehow related to the intensity of efforts required to ensure compliance of CS in the postgame phase.

Even more interesting is the result we obtain if CS moves first. Again, it would be quite unrealistic to imagine that CS, while moving first, would also be making false promises (as that would rather go against the logic). So what if the LG disagreed on the direction proposed by CS with its first move? Only in that case, there would be an incentive to cheat – or fight against the proposed outcomes. We can then conclude that while $U(RCS_{11})>U(RCS_{21})$ with good certainty, the relation between $U(RLG_{11})$ and $U(RLG_{12})$ can be any, thus making the final equilibrium indeterminate.

The lessons learned from this exercise are threefold:

1. The only way CS could condition the outcomes of the game is by "punishing" the LG that would like to run a real participatory game after previous cheating behavior. Thus, there is no room for an alternate fair/unfair approach to participation in (especially local) governments.
2. It is extremely unlikely that the LG would make room for bottom-up participation unless already aligned in its "philosophical inspiration" to CS activism (in which case, "bottom-up" initiatives can simply be seen in continuity with "top-down" ones).
3. Even in the case the LG had interest in inducing some kind of behavioral change in the CS representatives, bottom-up participation may not be acceptable to it (or aligned with its specific or generic goals).

In order to overcome some of the problems above, let us take into consideration a variant of the above scenario, whereby the rules of the game are no longer preexisting to the players and impossible to modify. In fact, when we speak about codesign of public services between LG and CS, we somehow describe a situation in which it would be quite unlikely that any party did not find some rationale or justification for their participation in the jointly defined (if not negotiated) game.

Although this is hardly the case in real life, where either the LG or CS continues to move first in carrying forward experiments (or proofs of concept) for the new and innovative services they envisage, we can keep valid as a first approximation the case where both parties first agree on the rules of cocreation and then start acting accordingly.

By so doing, they somehow lift the veil of ignorance that characterized the previous game. Both parties mutually recognize each other, accept the identification of a new set of rules, and comply with them. This bears the important consequence that the rule of law must contribute to outlining the most appropriate pathways for user integration in the public service delivery process and value chain.

7.6 Conclusions and Way Forward

Nowadays governments around the world are reconsidering efficiency, effectiveness, and quality of service delivery under the pressure of fiscal restraints, social demands, and reputational factors. However, most of the emerging and existing community challenges have little possibility (if any at all) of being successfully tackled unless the engagement and proactivity of large masses of citizens is won. This makes both the tactics of doing "more with more" and "more with the same" budget largely inappropriate.

On the one hand, the provision of high-quality services in a uniform way and at virtually no cost for the beneficiary has proven unaffordable, even for the most affluent governments and societies. People do want more and what they ask for is differentiation and customization of services – which is impractical from any point of view.

On the other hand, the logic of downsizing in order to save resources only creates monsters, with most of the budget cuts located at operational, rather than infrastructural, level, which leaves the bureaucratic burden virtually untouched and makes fewer and fewer customers happy with a fast declining quality of service, despite the generous efforts of frontline staff.

A new way of thinking is required to successfully manage change, which is called wise and frugal government. It is frugal because it is parsimonious in the use of resources, not only financial but also particularly human, adopting the diligence of a good family man in finding, measuring, and distributing the necessary and available resources. It is wise because it aims to empower rather than prevent the activism of citizens in the direction of societal transformation, to minimize the cost and maximize the benefit of public service delivery, thanks to its value-centered and user-driven perspective.

This new way of thinking is nailed on a renewed attention on the potential of "true" participation, namely, large-scale collaboration between (local) governments and civil society, to promote the generation of relational capital and, through this, ignite individual and collective learning and behavioral change.

Although some early examples of managerial "cookbooks" have started to appear, new sets of rules, binding for both sides of the frugal government – active citizenship relation – are to be designed and developed, in order to meet the promises of societal transformation as required by local and global challenges.

References

Accenture (2008) Global cities forum: exploring people's perspectives on the role of government. Institute of Public Service Value. New York. Available at: https://www.accenture.com/us-en/~/media/Accenture/Conversion-Assets/DotCom/Documents/Global/PDF/Industries_16/Accenture-Exploring-Perspectives-in-Government.pdf

Afonso A, Schuknecht L, Tanzi V (2003) Public sector efficiency: an international comparison. European Central Bank working paper no. 242. Frankfurt am Main, Germany. Available at: https://www.ecb.europa.eu/pub/pdf/scpwps/ecbwp242.pdf
Afonso A, Romero A, Monsalve E (2013) Public sector efficiency: evidence for Latin America. Inter-American Development Bank discussion paper no. IDB-DP-279. Frankfurt am Main, Germany. Available at: https://www.ecb.europa.eu/pub/pdf/scpwps/ecbwp242.pdf
Agranoff R, McGuire M (2004) Collaborative public management: new strategies for local governments. Georgetown University Press, Washington, DC
Allan JA (2011) The role of those who produce food and trade it in using and 'trading' embedded water: what are the impacts and who benefits? In: Hoekstra AY, Aldaya MM, Avril B (eds) Proceedings of the ESF strategic workshop on "Accounting for water scarcity and pollution in the rules of international trade", Value of water research report series no. 5431-52. UNESCO-IHE, Amsterdam, 25–26 November 2010
Bandura A (1977) Social learning theory. Prentice Hall, Englewood Cliffs
Beer A, Clower T (2014) Mobilizing leadership in cities and regions. Reg Stud Reg Sci 1(1):5–20
Bommert B (2010) Collaborative innovation in the public sector. Int Publ Manag Rev 11(1):15–33
Büttgen M, Ates Z (2009) Customer participation and its effects on service organisations: an institutional economics perspective. In: Proceedings of the 2009 Naples Forum on Service, Capri, Italy, June 16–19
Churchman CW (1967) Wicked problems. Manag Sci 14(4):B-141–B-142
Concilio G, Molinari F (2014) Urban living labs: learning environments for collective behavioural change. In: Proceedings of the IFKAD 2014 conference, 11–13 June, Matera, Italy
Concilio G, Marsh J, Molinari F, Rizzo F (2015) Human smart cities. A new vision for redesigning urban community and citizen's life. In: Proceedings of the 8th international conference on Knowledge, Information and Creativity Support Systems (KICSS'2013), 7–9 November, Krakow, Poland
Cooper A (2004) The inmates are running the Asylum: why high-tech products drive us crazy and how to restore the sanity. Sams Publishing Co, Indianapolis
CS Transform (2010) Citizen service transformation: a manifesto for change in the delivery of public services. http://www.cstransform.com/resources/white_papers/CitizenServiceTransformationV1.pdf. Accessed June 2015
Dell'Era C, Landoni P (2014) Living lab: a methodology between user-centred design and participatory design. Creat Innov Manag 23(2):137–154
Denhardt RB, Denhardt JV (2000) The new public service: serving rather than steering. Public Adm Rev 60(6):549–559
Döös M, Wilhelmson L (2011) Collective learning: interaction and a shared action arena. J Work Learn 23(8):487–500
Dunleavy P, Hood C (1994) From old public administration to new public management. Publ Money Manag 14(3):9–16
Eskelinen J, Garcia Robles A, Lindy I, Marsh J, Muente-Kunigami A (2015) Citizen-driven innovation: a guidebook for City Mayors and Public Administrators. World Bank and Brussels, European Network of Living Labs, Washington, DC, https://openknowledge.worldbank.org/bitstream/handle/10986/21984/Citizen_Driven_Innovation_Full.pdf?sequence=9. Accessed Aug 2015
European Commission (2010) eGovernment action plan 2011–2015. http://eur-lex.europa.eu/LexUriServ/LexUriServ.do?uri=COM:2010:0743:FIN:EN:PDF, Accessed Aug 2015
European Commission, DG CNECT (2013) A vision for public services. http://ec.europa.eu/newsroom/dae/document.cfm?doc_id=3179. Accessed Aug 2015
Fleming L (2004) Perfecting cross-pollination. Harv Bus Rev 82(9):22–24
Fuchs VR (1968) The service economy. Columbia University Press, New York
Graham F (2010) M-Pesa: Kenya's mobile wallet revolution. BBC News. http://www.bbc.co.uk/news/business-11793290. Accessed June 2015

Harrison T, Pardo T, Cresswell A, Cook M (2011) Delivering public value through open government. The Research Foundation of State University of New York. Center for Technology in Government. University at Albany. http://www.ctg.albany.edu/publications/issuebriefs/opengov_pubvalue.pdf. Accessed Sept 2015

Hood C (1991) A public management for all seasons? Public Adm 69:3–19

Hood C (1995) The new public management in the 1980s; variations on a theme. Acc Organ Soc 20(2/3):95–109

Hui G, Hayllar MR (2010) Creating public value in E-Government: a public-private-citizen collaboration framework in Web 2.0. Aust J Publ Admin 69:S120–S131

Immelt JR, Govindarajan V, Trimble C (2009) How GE is disrupting itself. Harv Bus Rev 87(10):56–65

Jefferson T (1801) First inaugural address as 3rd President of the United States in Washington, DC, March 4th, 1801. http://www.bartleby.com/124/pres16.html. Accessed Apr 2015

Krafcik JF (1988) Triumph of the lean production system. Sloan Manag Rev 30(1):41–52

Mansuri G, Vijayendra R (2013) Localizing development: does participation work? World Bank, Washington, DC

Marsh J (ed) et al (2014) The human smart cities cookbook. http://issuu.com/planumnet/docs/cookbook_planum. Accessed Sept 2015.

Molinari F (ed) et al (2013) How to set up cross border living labs. The alcotra innovation experience handbook. The Alcotra Innovation Project Consortium. https://www.alcotra-innovation.eu/livingLabs/dwd/Alcotra_Innovazione_Handbook_2013.pdf. Accessed June 2015

Neirotti P, De Marco A, Cagliano AC, Mangano G, Scorrano F (2014) Current trends in Smart City initiatives: some stylised facts. Cities 38:25–36

Orszag PR (2009) Open government directive. Memorandum for the heads of executive departments and agencies. Executive Office of the President, Office of Management and Budget, Washington, DC

Pallot M, Trousse B, Senach B, Scapin D (2010) Living lab research landscape: from user centred design and user experience towards user cocreation. First European Summer School "Living Labs", Paris

Rittel HWJ, Webber MM (1973) Dilemmas in a general theory of planning. Policy Sci 4:155–169

Siemens G (2005) Connectivism: a learning theory for the digital age. Int J Instr Technol Dist Learn 2(1). http://www.itdl.org/journal/jan_05/article01.htm. Accessed Aug 2015

Singh SK, Gambhir A, Sotiropoulos A, Duckworth S (2012) Frugal innovation. Learning from social entrepreneurs in India. The Serco Institute, London

Skidmore P, Bound K, Lownsbrough H (2006) Community participation. Who benefits? DEMOS and the Joseph Rowntree Foundation, UK

Surowiecki J (2004) The wisdom of crowds. Doubleday, New York

Svallfors S (2013) Government quality, egalitarianism, and attitudes to taxes and social spending: a European comparison. Eur Polit Sci Rev 5(3):363–380

Szkuta K, Osimo D, Pizzicanella R (2012) When people meet data. Collaborative approaches to public sector innovation. In: Proceedings of the 1st international EIBURS-TAIPS conference on "Innovation in the public sector and the development of e-services", University of Urbino, Italy, April 19–20

Tata Motors (2015) A promise is a promise. The story of the Nano. http://www.tatanano.com/nano-story.html. Accessed June 2015

W.K. Kellogg Foundation (1998, 2004). Using logic models to bring together planning, evaluation, and action. Logic model development guide. https://www.wkkf.org/resource-directory/resource/2006/02/wk-kellogg-foundation-logic-model-development-guide. Accessed June 2015

Chapter 8
From Public-Private-People Partnerships to Trading Zones in Urban Planning

Raine Mäntysalo

Abstract The chapter reviews critically the complexities involved in the idea of Public-Private-People Partnership (4P) in urban planning. The 4P idea, formulated by Majamaa and his colleagues, reaffirms the domination of the economic communication mode, familiar from Public-Private Partnerships, despite its aim of engaging the "people" in the partnerships. It thus undermines considerations on, e.g., political accountability, legal status, and scientific validity that may emerge in such partnership arrangements. Following Luhmann's social theory, each of these considerations has its own rationale that cannot be subjected to the economic mode of communication. The political, legal, and scientific considerations stem from their own communication modes, each with their own rules. This raises the question of how the coexistence of the different communication modes can be managed in 4Ps, without losing the partnerships' capability to perform. In discussing this challenge, the chapter focuses on the possibility of co-coordination between the communication modes, through the development of tools and platforms for boundary-crossing communication. For this purpose, the concepts of "boundary object" and "trading zone" are examined. Sociological studies in the history of science, by Galison and others, have revealed that different groups of scientists and experts have been able to co-coordinate their activities locally, by developing boundary objects and trading zones for exchanging information and services – despite not sharing their goals and conceptual understandings. A tentative analogy is drawn to "urban living labs." They are conceived as local semifixed platforms that combine spatial facilities and mapping, monitoring, and visualization technologies for the development of boundary objects and trading zones – in co-coordinating different views and understandings on urban planning issues. In this regard, an innate resource of planning is its storytelling approach to communication. In the conclusion, a reconceptualization of the 4P in urban planning is suggested, as a local trading zone of urban planning. Finally, some issues requiring further theoretical work are addressed.

R. Mäntysalo (✉)
Department of Built Environment, Aalto University, Espoo, Finland
e-mail: raine.mantysalo@aalto.fi

G. Concilio, F. Rizzo (eds.), *Human Smart Cities*, Urban and Landscape Perspectives 19,
DOI 10.1007/978-3-319-33024-2_8

Keywords Boundary object • Function system • Institutional ambiguity • Urban living lab

8.1 Introduction

Public-Private Partnerships of different sorts have been commonplace in urban planning for several decades. More recently, the concept of *Public-Private-People Partnership* has been introduced to add responsiveness of the partnership to the citizens' needs. However, with the addition of "people" to the partnership, further complexities arise, while some critical issues (e.g., public accountability) remain. This requires further theoretical work.

The aim of this chapter is to examine the theoretical aspects of the Public-Private-People Partnership (4P) idea, especially in reference to urban planning in the Finnish context. The study starts by recalling the development of Public-Private Partnerships in urban planning and then discusses the criticism they have received. The 4P model is discussed as a partial response to this criticism. First, the 4P model by Majamaa et al. (2008) is introduced. Then the narrowness of its approach to local governance is revealed in the broader context of different coexisting modes of local governance. As a potential response to this criticism, the approach of Kuronen (2011) is then scrutinized, as it conceives the 4P as a constellation of different mutually interacting systems. However, as argued below, a more elaborate systems perspective to 4P is needed.

For this end, the *Luhmannian systems approach* is introduced. In his sociological theory, Luhmann conceives modern societies to be constituted of different coexisting function systems that reproduce themselves by communicating in terms of their own distinctive codes and laws. Each function system, such as politics, economics, law, and science, has its own distinctive communication mode that is structured on the basis of its own core distinction, whereby it separates itself from its environment, e.g., government/opposition, ownership/non-ownership, legal/illegal, and true/false.

The Luhmannian approach raises the question of how these different function systems can be mutually coordinated. This is a question that was less considered by Luhmann himself, as he focused more on elaborating the codes and rules of each separate function system. He addressed the issue in his study of ecological communication in modern society (Luhmann 2004). Since, according to Luhmann, ecological communication does not have a corresponding societal function system, ecological goals have to be promoted "indirectly" through the coding mechanisms of relevant function systems that codify ecological issues on their own terms. Thereby the ecological issue appears as a political, economic, legal, and scientific (or other) issue. Striving for an ecological goal would thus entail the simultaneous harmonization of the relevant function systems' own ends – so that the ecological goal would make sense as a political, economic, legal, and scientific (or other) goal, as well.

The 4P concept aims to provide a methodology for the public, private, and "people" partners' cooperation in terms of a joint partnership. Now, if such a partnership is revealed to have tensions and difficulties, due to the involvement of different communication modes, then the problem is how a more "sensitive" cooperation methodology could be developed that would enable the mutual harmonization of the different goals that emerge with the different communication modes employed by the actors. In this chapter, the focus is on urban planning in particular. Similarly to Luhmann's ecological communication, urban planning can be said not to belong to any societal function system specifically, but to emerge as a societal issue as encoded into an issue having political, economic (and other) implications, each in turn. Then, in order to be successful, urban planning too would entail the mutual harmonization of the different political, economic (and other) goals that emerge when the involved actors "translate" urban planning issues in terms of their respective communication modes.

In theoretically examining the possibilities and methods of harmonizing different coexisting communication modes, Star and Griesemer's (1989) concept of *boundary object* and Galison's (1997) concept of *trading zone* are discussed. When coining these concepts, Star and Griesemer (as well as Galison) were interested in studying the conditions that in certain historical settings had enabled the mutual coordination of actors that, in Star and Griesemer's terms, belonged to different "social worlds." According to Star and Griesemer (1989), such objects can be taken into use, or purposely designed, so that they are robust and flexible enough to be used across the boundaries of these different social worlds – as tools for sharing or storing information or as shared objects of action – while, at the same time, these objects would permit their use in the inner, more elaborate operations of each social world, respectively. A few key boundary objects can be combined to provide a local platform for the co-coordination of actors that, while occupying different social worlds, are mutually dependent and stand to gain from reaching the ability to cooperate. Galison built on Star and Griesemer's work, focusing on the emergence and evolution of such local platforms, especially regarding "multicultural" cooperation in the different realms of science, such as nanotechnology and biochemistry.

In this chapter, the trading zone concept is examined as a potential answer to the problem of coordinating different communication modes in the context of urban planning. Can urban planning be conceived as a local platform for coordinating the stakeholders' different communication modes, utilizing such interlinguistic potentialities that characterize trading zones? As we shall see, certain aspects of urban planning have "interlinguistic" power in their capability to communicate across the boundaries of the stakeholders' different "social worlds" or "communication modes."

8.2 Public-Private Partnership

Since the rise of neoliberal political ideology in the late 1970s, first in the USA and the UK, the redistributive welfarism of public planning has been under attack as bureaucratic, inefficient, and reactive. In the aftermath of the 1970s recessions, public planning was expected to actively support the economy and the functionality of the globalizing markets. In urban planning, instead of guarding the public interest in a regulative manner, municipalities were expected to smoothen planning procedures and regulations and engage in private land development initiatives as a potential public partner and/or provider of different sorts of incentives (see a comprehensive account by Sager 2011).

The Nordic countries have also been heavily influenced by the political ideology of neoliberalism, with the associated reorganization of governmental structures in terms of new public management (NPM). The bureaucratic ethos has increasingly been replaced with operating principles drawn from the private sector, such as competitive bidding, outsourcing, purchaser-provider models, and Public-Private Partnerships. The mode of governance has received traits of *managerialism*. It approaches the municipality as a corporation, offering the municipal council a role resembling a managing board that makes the strategic decisions and hands their operational implementation over to the public officials (e.g., Möttönen 1997). Citizens are seen as customers of municipal services. Roivainen (2002) calls this mode of local governance *service municipality*. It aims at customer satisfaction, and the planning professional is no longer expected to plan for the universal citizen. He/she plans for the customer and for the desired customer especially. It is seen to be in the public interest to attract taxpayers and investors and thereby to promote local growth and competitiveness. What is "good" planning is seen to be identifiable through market behavior.

Regarding urban planning, the key policy tools of NPM are *Public-Private Partnerships*, with market actors having an active role in detailed planning and development projects (see Sager 2011; Mäntysalo 1999; Kurunmäki 2005). Public-Private Partnerships are an American innovation. In the USA they have been perceived as the key for urban revitalization, as the federal revenues for economic development, welfare services, and other urban programs have diminished (Squires 1996: 266). Squires, however, argues that the Public-Private Partnership in the USA is merely a newer name for a long-standing close relationship between private firms and public agencies (Squires 1996: 267). Public-Private Partnerships take many forms (Squires 1996: 266–267; Healey 1997: 267; Mayer 1997: 237–238; Sager 2011). Some partnerships are formal organizations, others informal cooperation arrangements. Some have persisted for decades, working with a multitude of issues, while some others are ad hoc arrangements that focus on a certain project, limited in both time and space. What they have in common is that the private sector receives direct subsidies from the public sector (Squires 1996: 266–267). Among the typical projects are those in which private housing, offices, commercial centers, and recreational facilities are designed for abandoned industrial sites and harbor areas situated close to the city center, which conceal land value development potential. Beyond the actual site, the project usually covers the development of public spaces and infrastructure, as well (e.g., Sager 2011).

In urban planning, the Public-Private Partnership often follows the logic of the plus-sum game, where every party gains a surplus for its investment, be it money or land property. Public-Private Partnerships vary greatly in their openness and responsiveness to local interests, depending on local political traditions and the current balances of power (Mayer 1997: 239). The partnerships that focus on growth-promising central areas especially often involve an exclusive inner circle that represents selected interests only.

In Finland, Public-Private Partnerships have also become popular in urban planning. As tax revenues and national subsidies have decreased, the competition between municipalities for private investment and taxpayers has also intensified, especially in the urban regions comprising several municipalities (Hytönen et al. 2012, 2013). Following the principle of outsourcing and the associated purchaser-provider model, the Finnish municipalities are increasingly taking part in urban development partnerships of varying sizes. This usually includes a specially administered project plan, based on a private developer's or landowner's initiative. In economically less certain development projects, the municipality may share the risks involved with the private developer. Together with the developer side (the bank, the contractor, the state-managed organization, etc.), it forms a partnership that commits itself to prompt realization of the project.

A typical form of Finnish Public-Private Partnership is the *land use agreement*, concerning the planning and development of an area that is usually privately owned. In a sense, an argument similar to Squires' can be made regarding land use agreements in Finland. As noted above, Squires claims "Public-Private Partnership" to be a more recent name for a long-standing relationship in the US context. In a similar vein, although land use agreement was introduced as a legal instrument as late as 1999 in the Land Use and Building Act, land use agreements became popular already in the 1960s, when Finland became rapidly urbanized, and the rural municipalities neighboring Helsinki especially (e.g., Espoo and Vantaa) were faced with a rapidly growing population. To cope with this growth, the municipalities resorted to land use agreements with the developer-contractors that started to systematically purchase rural land for housing development (Hankonen 1994; Hirvonen-Kantola and Mäntysalo 2014). The basic motivating forces pushing both parties (the developer-contractor and the local government) toward a mutual land use agreement are still the same today. The developer has to depend on the local government's sharing of its aims, as in Finland the local governments are in charge of local land use planning, as they have the so-called planning monopoly. On the other hand, the local government is motivated to reach an agreement with the developer, as it needs the latter's financial sharing of costs for planning and infrastructure (Mäntysalo and Saglie 2010).

A survey of Public-Private Partnerships in Finland was published in 2007, including background data of 43 contemporary cases and deeper analyses of 13 cases (JYMY 2008). The findings of the survey stress the importance of early commitments between the partners to the shared project, in order to minimize the risks and to safeguard the project's smooth progress. In projects such as housing, where the land is usually owned by the private developer, the local government tends to conform to the developer's goals in broad terms. Local governance cultures can differ considerably and, accordingly, the agreement policies vary as well. Written agreement documents are not often used (JYMY 2008; see also Hakkola 2007).

Especially in the early stages of the project, it is commonplace to agree merely on a "handshake" basis, although in some cases a written preliminary agreement has been drawn up, describing common goals and programming of the project. In most cases, the first and only written agreement is the actual land use agreement devised just before the municipal council's approval of the finished detailed plan for the area (JYMY 2008.)

There are also local differences in the division of roles and responsibilities in planning. In large municipalities with sufficient planning resources, the respective detailed planning is conducted by the municipal planning agency, while the private developer is in charge of the area's preliminary planning, in addition to the project implementation design. In small municipalities the actual preparation of the detailed plan may be handed to a private planning consultancy, selected by the municipality or the developer. In these cases, too, the municipal planner's responsibility is to supervise the planning process and coordinate the associated procedures of public hearings, participation, and assessment. Detailed planning and implementation design are usually interlinked. This has been seen to aid in safeguarding developer commitment to the qualitative demands regarding the implementation. On the other hand, there have been difficulties in convincing the public, in the participatory detailed planning process, that planning goals have not been agreed upon beforehand (JYMY 2008; see also Mäntysalo and Saglie 2010).

8.3 4P: Public-Private-People Partnership

Public-Private Partnership planning is problematic from the point of view of the Nordic planning tradition with an emphasis on participation, openness, and the strong action of the local government. On the other hand, the often privately initiated partnership projects require early investments from the private developer, and the management of the risks involved obviously requires some sort of safeguarding through public-private agreements and commitments on the goals of the respective future plan. With the agreements, private law steps in, justifying secrecy on agreed issues, such as financial commitments, duties in implementation, and sanctions for failure to comply with the contract. A severe contradiction with the planning law's principles of participatory planning may follow, if the local government resorts to drafting agreements and planning schemes with the developers regarding the contents of the future plans before starting the official participatory planning processes. The local government may adopt an inherently contradictory role in partnerships where market criteria dominate. As an investor in the partnership, it shares the interest of economic benefit with the private partners, and it is tempted to use secretive strategies in its own planning. On the other hand, as a democratically governed public organization, it is assumed to guarantee public accountability of the plans it produces. It has difficulties in finding an inner balance to discern what ought to be treated as business secrets and what should be considered as public matters (Mäntysalo 1999; Mäntysalo and Saglie 2010; Mäkinen 2000).

The criticisms on Public-Private Partnerships, regarding the lack of public accountability and the loss of responsiveness to citizens' needs, have been raised internationally, not merely in the Nordic context (e.g., Puerari 2014). While acknowledging that Public-Private Partnerships are inevitable under the conditions of global capitalism, Fainstein argues that "what needs to be done is ensure that the public component is more controlling and shares more in the proceeds" (Fainstein 1997: 140).

Reflecting on the criticism regarding the loss of responsiveness to citizens' needs, Majamaa et al. (2008) have proposed the model of *Public-Private-People Partnership* (4P) for public service provision involving real estate development. They examine Public-Private Partnerships within the context of consumer society, where societal relations are seen to be dominated by consumerism. The Public-Private Partnership is approached from the point of view of the purchaser-provider model. The public body (e.g., municipality) is in charge of the public service for its citizens (perceived as end-user customers), and the private actor is contracted to provide this service to the public body. Majamaa et al. (2008) argue that in such settings, the focus of the partnership is on the interface between the public and the private actor and not on the actual end users (citizens) of the produced public service. There is no incentive for either party to develop the service based on the end users' actual feedback, as the public body focuses on following the legal requirements on the level of the public service and the private provider is looking for the cost-efficient provision of such service. End users are treated as homogeneous subjects of services which have no direct contact to the actual private service provider, but have to give their feedback to the public body responsible for the service via the local democracy channels.

Majamaa and his colleagues (2008) offer their 4P model in order to shift the focus to the end users (people) who are the "real" customers of the service – not the public purchaser. In a 4P setting, in addition to the formal local democracy channels with the public body, end users would have informal channels to influence private providers, which, in turn, would be encouraged to develop their service provision further – and even to create additional third-party services, in response to the end users' further needs related to real estate and facility development, thus exceeding the actual legal requirements on the public service in question. Thereby the 4P model would support active end-user participation in the production of public services, approaching the idea of *coproduction* (Leadbeater 2004; see also Wallin 2010). In his doctoral thesis, Majamaa (2008) extends this argument to urban planning and design, too, as forms of public service.

However, the 4P concept by Majamaa et al. (2008) can be criticized for its narrowness, regarding the tasks and duties of the municipality in connection to the provision of public services. Their concept of municipality corresponds to Roivainen's "service municipality" described above. But this is only one of the three coexisting modes of local governance that Roivainen (2002) has identified in the Finnish context. The other two are *administrator municipality* and *citizen municipality*.

The *administrator municipality* represents the traditional bureaucratic public sector. Decisions are largely determined within the administration, to be more or less rubber-stamped by the elected politicians. The formal authority of the municipal administrator derives from her/his position as a holder of public office, regulated by laws and rules of conduct. The professional's knowledge reflects the modernist belief in truth identifiable through scientific methods. The impartial and rational professional makes decisions based on facts. The citizens have a fairly passive role in this model: they are subordinates to the top-down order. Municipal politicians form a link between the citizens and the government: citizens vote for the representatives of their particular interests, and the representatives then communicate these interests in municipal decision-making (Roivainen 2002).

The *citizen municipality* embraces the idea of active citizenship. Citizens participate in the co-governance of their municipality through deliberating with public officers and politicians on collective issues. In this model, the administrator-officer has become a networker, a facilitator of interaction. The values that are forwarded in this model are participation, transparency, and public accountability (Table 8.1).

The neoliberal political ideology fosters the identification of local governance in terms of service municipality, while the coexisting civil society movement supports citizen municipality. On the other hand, the institutional framework of Finnish local governance continues to divide roles and duties in local administration and decision-making in accordance to the administrator municipality model, with its dualistic separation of the administrators' allegedly fact-based preparation of decisions and the politicians' value-based making of decisions.

Regarding planning, Bäcklund and Mäntysalo (2010) argue that this coexistence of different modes of local governance is a potential source of *institutional ambiguity* (concept borrowed from Hajer 2006):

> [W]e find ourselves in a complex reality of planning when different understandings of the determinants of good democracy, legitimate planning and roles of different actors in producing and managing valid knowledge coexist and compete with each other [. . .] This poses a critical challenge that has been largely ignored by the researchers in participatory planning. (Bäcklund and Mäntysalo 2010: 348)

The 4P model based on mere service relations (purchaser, provider, and customer of public services) is not capable of sorting out this institutional ambiguity. There are other kinds of relations, too, that stem from the administrator and citizen municipality modes of governance. As seen above, Majamaa et al. (2008) included local democracy as a formal channel for the citizens' customer feedback. This is

Table 8.1 Roles of actors in administrator, service, and citizen municipalities

	Administrator municipality	Service municipality	Citizen municipality
Public administrator	Executor	Expert	Networker
Elected politician	Representative	Manager	Interpreter of community
Citizen	Subject	Customer	Citizen

a strikingly narrow understanding of the nature of democracy. The introduction of People to the partnership model makes the picture much more complicated than merely adding new service relations.

In his doctoral thesis, Kuronen (2011; see also Kuronen et al. 2010) has attempted to develop the 4P model further. He approaches the relationships between "public," "private," and "people" in terms of three different subsystems: the *economic subsystem* set between the private and the people, the *administrative subsystem* between the public and the private, and the *political subsystem* between the people and the public (Kuronen 2011: 42). The approach is inspired by Mäntysalo's (2000) description of land use planning as a political system consisting of three subsystems, i.e., politics, expertise, and economics.

However, Kuronen's (and his colleagues') systems approach is not elaborate enough to grasp the differences between the three subsystems and the interrelations between them (see Joutsiniemi 2013). Mäntysalo's (2000) theoretical analysis of the land use planning system was focused on the contradictions and double binds within and between the subsystems of planning and on (inter)organizational learning in overcoming these contradictions and double binds while acknowledging the unavoidable persistence of tensions between the mutually interdependent subsystems. His systems approach was based on the combination of Luhmann's (1990) theory of autopoietic social systems and Bateson's (1987) cybernetic explanation of human behavior.

8.4 The Luhmannian Approach: Toward Co-coordination

The Luhmannian (Luhmann 1990, 2004) systems view approaches social systems as different *modes of communication*. These systems do not consist of actors but *roles* that the actors receive according to the systems' basic distinctions and programs. Each social system has its own basic distinction that encodes its communication, such as *ownership/non-ownership* in the system of economics, *government/opposition* in the system of politics, and *true/false* in the system of science. These distinctions also determine how actors enter these systems in terms of roles. An actor enters the economic system as an "owner" or "nonowner" (e.g., landowner, investor, buyer, consumer). Accordingly, the political system consists of the members and supporters of government or opposition, and the system of science consists of scientists and experts and laypersons.

Thus, if we analyze the Public-Private-People Partnership as interdependent coexistence of Luhmannian social systems, we are not actually dealing with public, private, and people as unambiguous actors any longer, but with the different roles that each of them receives in different systems that become actualized through the actors' mutual communication. Undoubtedly, in the urban planning context, economic, political, and scientific modes of communication are usually actualized, but others are as well, such as legal communication (*legal/illegal*), e.g., in the applications of the public and private law in planning work and in formalizing partnership relations.

In the Luhmannian view, Majamaa's and his colleagues' 4P model can be seen to follow essentially the communication mode of the economic system: the three-actor groups basically receive their roles according to the ownership/non-ownership distinction (purchaser, provider, customer). When the Public-Private-People Partnership is proposed as a normative model for the coproduction of public services, in the vein of Majamaa et al. (2008), we need to discern the economic communication mode at its origin. Yet, there are other communication modes (systems) that tend to become actualized in such partnership formation. The political justification of the partnership may be questioned. Who is involved and who is not in it? What is the public accountability of the partnership and its decisions? From the science perspective system in turn, what is the factual basis for the surveys made and plans proposed? Has the relevant knowledge been provided? Have proper experts been involved? From the law system point of view, the legal weight and consequences of the partnership are usually checked and may be tested, as well. Which other communication modes (systems) enter the scene, and how, is actually an empirical question to be examined in each case.

In a Luhmannian interpretation, the same criticisms that have been posed to the Public-Private Partnerships from a political perspective (e.g., lack of public accountability and openness, domination of market criteria) can be repeated with the 4P model, since, in essence, it extends the partnership merely in terms of economic communication. With its economic mode of communication, it cannot provide a political response to the criticisms that are made from the perspective of the political system, nor can it subject political communication to economic transactions. The political criticism can be addressed only if the political mode of communication is understood to coexist on its own terms, and then used, alongside the economic communication of the 4P. While the actors involved in the 4P are given economic roles, the political criticism reveals that they have political roles as well, like it or not. Who is the municipal councilor representing with his/her involvement in the partnership? Is the citizen, involved in the partnership, some kind of political representative and of whom? What is the status of the partnership in relation to the formal bodies of municipal decision-making? Political communication bears consequences on the economic system, and vice versa, but each can respond to these only in terms of their own codes and rules. The actors, however, are not victims to any system and corresponding role, but they can shift from one communication mode and role to another in pursuing their goals, and often do so with tactical skill (Fig. 8.1).

While the 4P model by Majamaa et al. oversimplifies the complexity related to 4P, Kuronen, in turn, although better addressing complexity with his systems approach, oversimplifies his view on systems, at least from the Luhmannian point of view. The Luhmannian idea of the coexistence of mutually interdependent, yet autopoietic (self-regulating), systems of modern society instructs us not to build any simple relationships and direct causalities between economics, politics, science, etc. Instead, if we wish to develop the 4P model as responsive to, e.g., political criticism, we need to focus on the *co-coordination of the different communication modes* that become actualized with it. Through co-coordination of different communication modes related to 4P, the challenge of institutional ambiguity can also be met.

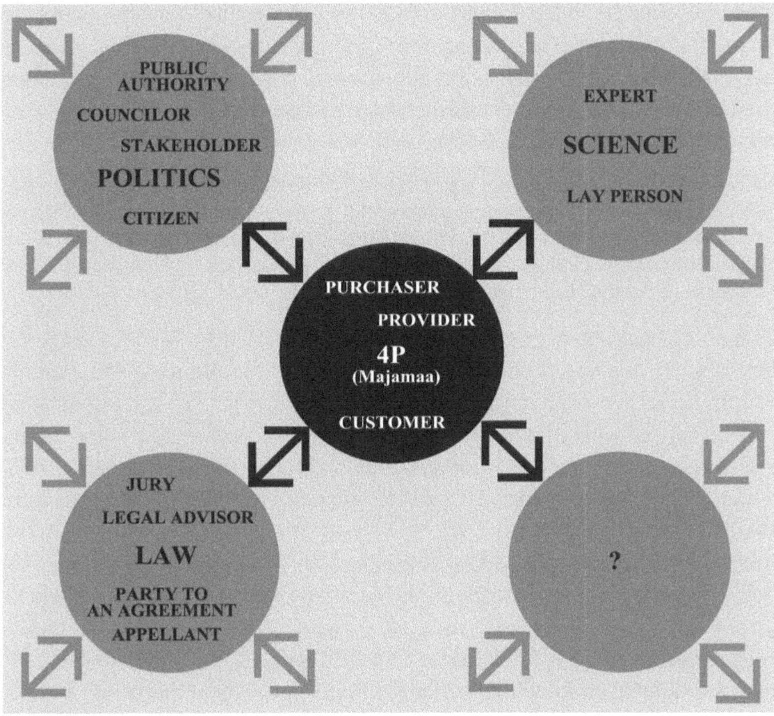

Fig. 8.1 The 4P model by Majamaa et al. (2008) with its actor roles, interpreted as an economic communication mode interacting with its environment of other communication modes (political, scientific, legal, etc.), in each of which the actors receive different roles

8.5 Boundary Objects, Trading Zones, and Urban Living Labs

In the search for means to enable the co-coordination of different communication modes, the concepts of *boundary object* and *trading zone* may be of use. These concepts have been coined as analytical aid in studies seeking to explain how different groups, such as scientists and policy-makers, have been able to develop, in certain specific circumstances, arrangements of coordinated interaction, despite the separateness of their "social worlds" (Star and Griesemer 1989), including separateness of basic goals and values and conceptual understandings.

This line of research stems largely from Star and Griesemer's groundbreaking case study of the establishment of the Berkeley's Museum of Vertebrate Zoology in the early twentieth century, in which they introduced the concept of boundary object (Star and Griesemer 1989). Star and Griesemer describe how the director of the newly founded zoological museum in the University of California, Joseph Grinnell, managed to develop and use a repertoire of instruments and objects when coordinating the activities of various actors with different motives and understandings, such as researchers, sponsors, university management, amateur collectors, and hunters. These instruments and objects included repositories, ideal types, standardized forms, and coincident (geographical) boundaries. According to Star and Griesemer (1989), they had the character of boundary object, since they could be used in a coordinated way as shared objects and tools of activity across different social worlds. In their definition:

> Boundary objects are objects which are both plastic enough to adapt to local needs and the constraints of the several parties employing them, yet robust enough to maintain a common identity across sites. [. . .] They have different meanings in different social worlds but their structure is common enough to more than one world to make them recognizable, a means of translation. (Star and Griesemer 1989: 393)

Star and Griesemer argue, further, that the "creation and management of boundary objects is a key process in developing and maintaining coherence across intersecting social worlds" (Star and Griesemer 1989: 393). Boundary objects constitute a sort of shared platform or infrastructure for coordinated interaction across different social worlds. For example, Harvey and Chrisman (1998) and Kahila-Tani (2013) have studied GIS technology as such a negotiated infrastructure between different social groups.

A related concept, introduced by Galison (1997), is the *trading zone*. Galison has studied interaction between theorists, experimentalists, and instrumentalists in particle physics, conceiving each as a subculture of its own. In accordance with Star and Griesemer, Galison has identified infrastructures of shared concepts and instruments that have enabled the exchange of information and services between the different "social worlds" of particle physics. Similarly to Star and Griesemer, in their reference to boundary objects, Galison stresses the *locality* of the trading zone: it is a specific site – partly symbolic, partly spatial – in which local coordination between theory and action takes place (Galison 1997). The concepts differ in their relation to development. Whereas boundary objects denote fixed infrastructures for mutual translation between different social worlds, trading zones refer to infrastructures that *evolve* and may go through different developmental changes, such as evolving from scientific jargon to pidgin and further to creole, a living hybrid interlanguage of science (see Galison 2010).

In their article, Mäntysalo et al. (2011) have outlined the so-called *trading zone approach* as a potential method for dealing with complex urban planning problems with multiple stakeholders. Combining the idea of trading zones and boundary objects with agonistic democracy, they seek to replace the consensus principle of the communicative planning theory with mutual coordination between the stakeholders. Galison's accounts of different scientists reaching the capability to cooperate successfully by intentionally *not* striving for consensus, but rather aiming at a sufficient level of co-coordination, provide a fresh alternative perspective to view the challenges of planning communication. The aforementioned article generated a book project (Balducci and Mäntysalo 2013) involving numerous case studies by Italian and Finnish researchers on trading zones in urban planning.

Leino (2008, 2012) has further studied the organizational aspects of participatory planning as boundary work. In sociological science and technology studies (STS), different kinds of institutional structures have been identified to emerge as facilitators of knowledge transfer between research, politics, and business. Some may be quite fixed "boundary organizations" (Guston 1999), while others may be "hybrids" that could rapidly change their form (Miller 2001). In urban planning we have witnessed the recent emergence of different *urban living labs* that may be perceived as some kind of boundary organizations. When such boundary organizations are generated in urban planning, it is crucial, from the point of view of political legitimacy, legal status, and scientific renewability, that they should not become too fixed and institutionalized. Continuous critical consideration needs to be maintained to prevent them from turning into exclusive clubs of selected stakeholders, which would thereby gradually lose their role as facilitators of communication and interaction, and challenge the formal organs of decision-making.

The concepts presented above suggest the possibility of approaching the 4P in urban planning as a semifixed local platform that could be developed into a trading zone between the stakeholders' different communication modes. In his groundbreaking study regarding co-coordination in microphysics, Galison noticed that certain practico-linguistic settings had been generated to enable the mutual exchange of knowledge and services between the scientists representing different "subcultures." Galison identified local infrastructures of shared concepts, laboratory equipment, and spatial settings that had facilitated such exchange. These infrastructures had functioned as platforms for the generation of localized "exchange languages." Such exchange languages had enabled the mutual "out-talk" between members of different subcultures, transforming highly elaborate and complicated issues into "thin descriptions." Accordingly, when a local urban living lab is developed with the aim of reaching trading zone quality, the focus would be on the whole practico-linguistic setting of the erected platform. It would focus on how the spatial arrangements of furnishing and equipping the meeting and studio room(s), the technologies of visualizing plans and monitoring development, and the verbal means of discussing planning issues would jointly contribute in creating the conditions for mutual "out-talk" on planning with "thin descriptions."

A specific feature that equips urban planning with such integrative power is its inclination to shape information in the format of *stories* (Throgmorton 1996; Forester 1999). The rhetorical strength of storytelling is its everyday familiarity. It involves certain scenes, different characters, and a plot with twists and turns that unfold with the story. A plan presented as a good story invites the listeners to share imagining the conditions, events, and episodes envisioned by it. Like a good history lesson, it concentrates on explaining the forces that influence the outcome of events, rather than plain numbers and names. In this way it is easier for people to react to the envisioned future and start discussing how to make it happen (Mäntysalo and Grišakov 2015).

Hence, planning as storytelling has trading zone characteristics. Without being limited to verbal means of communication, it provides tools for the mutual "out-talk" with "thin descriptions." In an urban living lab of urban planning, storytelling is a key means in generating a trading zone, enabling co-coordinative inquiry across the boundaries of different communication modes.

8.6 Conclusion

In this chapter, the complexities involved in the idea of Public-Private-People Partnership (4P) have been critically examined through the theoretical perspective offered by Luhmann's social theory. This perspective focuses on the differences between the different communication modes that become actualized in such partnerships. The economic communication mode that provides the rationale for Public-Private Partnerships in arranging public service relations has been further applied by Majamaa, who has proposed more "customer-sensitive" 4Ps in urban planning and development. His approach, however, undermines considerations on issues that may emerge in such partnership arrangements, such as political accountability, legal status, and scientific validity. Following Luhmann, each of these considerations has its own rationale that cannot be subjected to the economic mode of communication. Political, legal, and scientific considerations stem from their own communication modes, each with their own rules. This raises the question of how the coexistence of the different communication modes can be managed in 4Ps, without losing the partnerships' capability to perform.

This challenge calls for highly reflective leadership and orchestration, but also for the development of intermediating tools and platforms, for the co-coordination of different arguments and activities that stem from the different coexisting rationales. In this chapter the focus is on the latter. The concepts of "boundary object" and "trading zone" are fruitful in this regard. They have been coined, in the realm of sociological studies of science and technology (STS), to describe and analyze specific practico-linguistic capabilities that have been generated in certain local contexts. Different groups that do not share goals and conceptual understandings have been able to co-coordinate their activities by developing boundary objects and trading zones for the joint exchange of information and services. In a sense, specific local platforms for the generation of "win-win-win settings" have thus been created. In certain cases, such local platforms may have been developed into more or less permanent and institutionalized "boundary organizations."

In this chapter, a tentative analogy has been drawn to "urban living labs" that could be seen as semifixed local boundary organizations of 4P urban planning. As such, urban living labs would use spatial facilities, as well as mapping, monitoring, and visualization technologies, for the development of boundary objects and trading zones – in co-coordinating different views and understandings on urban planning. The inclination to communicate planning issues in terms of stories in particular, both verbally and visually, can be perceived as a resource that has innate trading zone qualities. Stories bring people together.

This line of reasoning suggests the reconceptualization of the 4P, in the context of urban planning, into an *urban planning living lab*, with the above described trading zone capacities.

If so, how should the "win-win-win settings," aimed in at such living labs, be understood? This question brings us to the difference between Star and Griesemer's concept of "social world" and Luhmann's concept of "communication mode." While Star and Griesemer seem to approach the different social worlds as if they were occupied by different "people," Luhmann's different communication modes are not occupied by "people" but by "roles." This is a more fundamental conceptual difference than might appear at first glance. The conditions on which the Luhmannian roles may enter into a win-win-win situation are more complex. As an individual person receives many roles simultaneously in different function systems, reaching a win-win-win settlement is also a question of a person settling with his/her own different coexisting roles (political, economic, etc.). The roles have different principles and obligations attached to them. It is thus not just a question of "should I agree?", but "am I entitled, in this role, to agree?" Further theoretical work is required to assess the possible limitations of treating "social worlds" and "communication modes" as analogous to each other, the way it has been done in this chapter.

Moreover, a too straightforward translation to the context of urban planning of the ideas developed in the realm of STS on trading zones and boundary objects might lead to oversimplifications. As noted by Healey (2014: 927):

> Work in the sociotechnical systems tradition is full of useful ideas about how to analyze the social interactions between disparate groups, but it is important to remember that the institutional settings in which urban governance and urban planning activity take place is much more complex than in the scientific community, with more complex ways in which knowledge claims and legitimacy are established, and a much greater diversity of groups involved in interactions.

References

Bäcklund P, Mäntysalo R (2010) Agonism and institutional ambiguity: ideas on democracy and the role of participation in the development of planning theory and practice – the case of Finland. Plan Theory 9(4):333–350

Balducci A, Mäntysalo R (eds) (2013) Urban planning as a trading zone. Springer, Dordrecht

Bateson G (1987, orig.1972). Steps to an ecology of mind. Jason Aronson, Northvale

Fainstein SS (1997) Urban redevelopment and public policy in London and New York. In: Healey P, Cameron S, Davoudi S, Graham S, Madani-Pour A (eds) Managing cities. The new urban context. Wiley, Chichester, pp 127–143

Forester J (1999) The deliberative practitioner. Encouraging participatory planning processes. MIT Press, Cambridge, MA

Galison P (1997) Image and logic: a material culture of microphysics. University of Chicago Press, Chicago

Galison P (2010) Trading with the enemy. In: Gorman ME (ed) Trading zones and interactional expertise. Creating new kinds of collaboration. MIT Press, Cambridge, MA, pp 25–52

Guston DH (1999) Stabilizing the boundary between US politics and science: the rôle of the office of technology transfer as a boundary organisation. Soc Stud Sci 29(1):87–112
Hajer MA (2006) The living institutions of the EU: analysing governance as performance. Perspect Eur Polit Soc 7(1):41–55
Hakkola E (2007) Osapuolten intressit hankkeeseen. In: Ahlava A, Edelman H (eds) Urban design management. Opas käytäntöön. DECOMB, Espoo
Hankonen J (1994) Lähiöt ja tehokkuuden yhteiskunta. Suunnittelujärjestelmän läpimurto suomalaisten asuntoalueiden rakentumisessa 1960-luvulla. Tampereen teknillinen korkeakoulu, arkkitehtuurin osasto 551. Gaudeamus/Otatieto, Tampere
Harvey F, Chrisman N (1998) Boundary objects and the social construction of GIS technology. Environ Plan A 30(9):1683–1694
Healey P (1997) Collaborative planning: shaping places in fragmented societies. Macmillan, London
Healey P (2014) Book review: urban planning as a trading zone. J Reg Sci 54(5):925–927
Hirvonen-Kantola S, Mäntysalo R (2014) The recent development of the finnish planning system – the city of Vantaa as an executor, fighter and independent actor. In: Reimer M, Getimis P, Blotevogel H (eds) Spatial planning systems and practices in Europe. A comparative perspective on continuity and changes. Routledge, London, pp 42–60
Hytönen J, Mäntysalo R, Akkila I, Kanninen V, Niemi P (2012) Kaupunkiseutujen kasvukivut II. Paras-ARTTU-ohjelman tutkimuksia nro 22, Acta nro 241, Kuntaliitto, Helsinki
Hytönen J, Mäntysalo R, Peltonen L, Kanninen V, Niemi P, Simanainen M (2013) Defensive routines in land use policy steering in Finnish urban regions. European Urban and Regional Studies, Published online ahead of print 3 July 2013, doi:10.1177/0969776413490424
Joutsiniemi A (2013) Mitä peetä? Yhdyskuntasuunnittelu 51(1):67–70
JYMY – Julkisen ja yksityisen sektorin yhteistyö maankäytössä (2008) Suomen Kuntaliitto, Helsinki. http://kuntaliitto.fi/intra/julkaisut/pdf/p081128120058M.pd/, Retrieved 14 June 2009
Kahila-Tani M (2013) SoftGIS development process as a trading zone: challenges in implementing a participatory planning support system. In: Balducci A, Mäntysalo R (eds) Urban planning as a trading zone. Springer, Dordrecht, pp 75–93
Kuronen M (2011) The role of partnerships in sustainable urban residential development. Aalto University publication series doctoral dissertations 63, Espoo
Kuronen M, Junnila S, Majamaa W, Niiranen I (2010) Public-private-people partnership as a way to reduce carbon dioxide emissions from residential development. Int J Strateg Prop Manag 14(3):200–216
Kurunmäki K (2005) Partnerships in urban planning. "Development Area" in National and Local Contexts in Finland, Germany and Britain. Tampere University of Technology, DATUTOP 26, Tampere
Leadbeater C (2004) Personalisation through participation. A new script for public services. Demos, London
Leino H (2008) Kansalaisosallistuminen kaupunkisuunnittelussa: rajaorganisaatioita vai hybridien hallintaa? Alue ja ympäristö 37(2):41–48
Leino H (2012) Boundary interaction in emerging scenes: two participatory planning cases from Finland. Plan Theory Pract 13(3):383–396
Luhmann N (1990) Political theory in the welfare state. de Gruyter, Berlin
Luhmann N (2004) Ekologinen kommunikaatio. Gaudeamus, Helsinki
Majamaa W (2008) The 4th P – people – in urban development based on public-private-people partnership. TKK structural engineering and building technology dissertations: 2 TKK-R-VK2, Espoo
Majamaa W, Junnila S, Hemanta D, Niemistö E (2008) End-user oriented public-private partnerships in real estate industry. Int J Strateg Prop Manag 12(1):1–17
Mäkinen E (2000) Maankäyttösopimus ja hyvä hallinto. Finnpublishers, Tampere
Mäntysalo R (1999) Learning from the UK: towards market-oriented land-use planning in Finland. Hous Theory Soc 16(4):179–191

Mäntysalo R (2000) Land-use planning as inter-organizational learning. Acta Universitatis Ouluensis, Technica C 155, Oulu
Mäntysalo R, Grišakov K (2015) Framing 'evidence' and scenario stories in strategic spatial planning. In: Albrechts L, Balducci A, Hillier J (eds) Situated practices of strategic planning. Routledge, London (forthcoming)
Mäntysalo R, Saglie I-L (2010) Private influence preceding public involvement: strategies for legitimizing preliminary partnership arrangements in urban housing planning in Norway and Finland. Plan Theory Pract 11(3):317–338
Mäntysalo R, Balducci A, Kangasoja JK (2011) Agonistic planning as communication in a trading zone. Plan Theory 10(3):257–272
Mayer M (1997) Urban governance in the post-fordist city. In: Healey P, Cameron S, Davoudi S, Graham S, Madani-Pour A (eds) Managing cities. The new urban context. Wiley, Chichester, pp 231–249
Miller C (2001) Hybrid management: boundary organisations, science policy, and environmental Governance in the climate regime. Sci Technol Hum Values 26(4):478–500
Möttönen S (1997) Tulosjohtaminen ja valta poliittisten päätöksentekijöiden ja viranhaltijoiden välisissä suhteissa. Suomen kuntaliitto, Helsinki
Puerari E (2014) Possible new governance models in the innovation of urban public services. Paper presented at the AESOP Congress, Utrecht, July 12, 2014
Roivainen I (2002) Täällä Kaino – kuuleeko kunta? Päättäjänä kuntalaisten ja virkamiesten välimaastossa. Janus 10(3):266–273
Sager T (2011) Neo-liberal urban planning policies: a literature survey 1990–2010. Prog Plan 76(4):147–199
Squires GD (1996) Partnership and the pursuit of private city. In: Fainstein SS, Campbell S (eds) Readings in urban theory. Blackwell, Malden, pp 266–290
Star S, Griesemer JR (1989) Institutional ecology, 'Translations' and boundary objects: amateurs and professionals in Berkeley's Museum of Vertebrate Zoology, 1907–39. Soc Stud Sci 19: 387–420
Throgmorton JA (1996) Planning as persuasive storytelling: the rhetorical construction of Chicago's electric future. University of Chicago Press, Chicago
Wallin S (2010) The co-evolvement in local development – from the triple to the quadruple helix model. Paper presented at Triple Helix VIII, Madrid, October 2010

Chapter 9
New Urban Services: Toward New Relations Between Economy and Society

Jean-Louis Laville

Abstract The concept of "smart city" insists on the importance of new urban services, and two key questions have been identified in this direction. The first concerns the interplay between urban forces and the urban and microlevel governance mechanisms, and the second concerns the interactions between physical structures and digital information infrastructures.

This essay tries to introduce a historical perspective, showing that the idea of new urban services has its roots in the evolution of economies from industry to services, which occurred in the late twentieth century and led to a new strategic position for local initiatives. The specific dynamics of such initiatives have to be understood through a comprehensive methodology, allowing them to conceive innovative policies at local level that generate and enlarge institutional diversity. When such conditions are met, social enterprises are able to improve well-being because they adopt a public dimension and a hybridization logic, as explained by the ideal type of solidarity-based services.

Difficulties have to be clearly faced: to provide a new synergy between public authorities and citizen initiatives requires not only technological tools but also a new perception of the socioeconomic and political dimensions of civil society, far beyond the usual sectorial approach. Some experiments as well as some theoretical currents in Europe have provided a number of resources to reframe the links between economy and society. They have to be mobilized in human smart cities. Such phenomena have to be understood in depth because the cooperation between citizens, experts, and elected representatives is not obvious and a harmonious development of contemporary cities calls for huge changes that this chapter tries to present both empirically and theoretically.

Keywords New urban services • Civic entrepreneurship • Social enterprises • Social economy

---

J.-L. Laville (✉)
Conservatoire National des Arts et Métiers, Paris, France
e-mail: jean-louis.laville@cnam.fr

G. Concilio, F. Rizzo (eds.), *Human Smart Cities*, Urban and Landscape Perspectives 19,
DOI 10.1007/978-3-319-33024-2_9

9.1 Local Initiatives and Social Enterprises

At a European level, the public interest for local initiatives began with the *White Book* dedicated to the challenges and ways forward into the twenty-first century (Commission of the European Communities 1994). It concluded that meeting new needs and creating new jobs was a key issue for the future of our European lifestyles. A survey was therefore conducted to look at this issue in greater depth and it produced a European Commission working document, *Local development and employment initiatives* (Commission of the European Communities 1995). The document corroborated the importance of new activities and pinpointed supply areas likely to meet new needs: day-to-day services, services to improve the quality of life, cultural and leisure services, and environmental services.

9.1.1 Local Initiatives: Collective Involvement and Civic Entrepreneurship

In general, local initiatives tend to give priority to the service relationship within economic activities. Although everyone agrees that new jobs will, in the future, come chiefly from the service sector, it should be borne in mind that the range of services covered by local initiatives is relatively independent from the industrial system and the supply of goods. It takes the form of specific relational services, such as services based on the direct interaction between the provider and the user, whether this is a result of the nature of the activity (e.g., health and personal care) or of operating choices (such as types of environmental protection geared toward making citizens more responsible).

There is therefore a clear dividing line between these services and other services, such as services that can be standardized (banks, insurance, telecommunications, etc.), which deal with information that can be readily encoded, can consequently follow an industrial-type path, and work toward substantial productivity gains.[1]

At a time when the industry and standardizable services that provided a basis for expansion up to the 1970s are faced both with the need to compete in international markets and with large-scale changes in information processing technologies (that prevent them from continuing to generate jobs as they have done in the past), local initiatives are therefore a *way of exploring a European channel for the creation of new relational service activities and jobs*. This is what makes them important and paves the way for a better understanding of their characteristics and the dynamic that leads to their creation.

[1]On the difference between relational services and services that can be standardized, see Baumol (1987) and Roustang (1987).

The first component of creation for local initiatives involves the types of mobilization generated by idea-to-project transition. What is striking about local initiatives, in comparison with other forms of enterprise creation, is their ability to gain backing from social support networks. Local initiatives are able to rally partners from different backgrounds.

While the profiles of the people involved are very disparate, the fact that their own experience has made them *demand-side stakeholders*[2] is a common feature. It is this common feature that enabled these support groups to genuinely invent services because their starting point was their implicit or explicit perception that there were no suitable answers to the problems that they were encountering. This approach differs from standard approaches shaped solely by market studies or needs analysis. Their local character is paramount in this respect since their creation is shaped by a dual notion of proximity, i.e., *objective proximity* shaped by a local area and *subjective proximity* shaped by the relational dimension of the activity.

Creating a local forum for dialogue based on interactive exchanges makes it possible to match supply and demand in a way that avoids imposing stereotyped answers on the identified needs. This kind of *local public space* helps to move away from the coproduction inherent in all services toward *a joint construction of supply and demand*, in which users play a key role, either through their direct initiative, through the intervention of professionals who have become aware of unsatisfied demands (because they are immersed in the local fabric), or through their association with other partners, who become concerned about the issue in question for personal reasons.

If social networks are active components of local initiatives, they need the indispensable catalyst of an entrepreneur if they are to be successful. It would seem, in this respect, that risk-taking is not just motivated by material interests. While financial considerations are undoubtedly important, they build new ways of "living together" through their shared concern for a common good. A feature of entrepreneurs is their desire to promote a sense of social responsibility at a local level through economic activity. It is for this reason that entrepreneurs who build up local initiatives can be called *civic entrepreneurs*, because their economic action is shaped by a model of society based on sustainable development and social justice.

9.1.2 From Local Initiatives to Social Enterprises

If local initiatives share common features in their emergence, the forms of institutionalization differ. The first one is the for-profit private enterprise that sells goods and services through the market. The second one is the public enterprise that uses public funds to meet general interests.

[2]Using the term "demand-side stakeholders" as put forward by A. Ben Ner, T. Van Hoomissen, 1991: 519–549.

These two forms are not surprising, but what is more surprising is the importance of a third form, in which monetary and nonmonetary resources are combined. It seems essential that social enterprises are not dominated by market or state logics.

Not only do the social enterprises jointly shape supply and demand in public spheres for local services, but they also consolidate their position by *combining monetary and nonmonetary resources.*

As far as nonmonetary resources are concerned, although they help the established structures to survive, they will certainly not be enough to guarantee the future of these structures, because activities which are too reliant on voluntary help tend to be nothing more than temporary phenomena dependent on favorable circumstances. For these reasons, hybridization between resources is a way of consolidating social enterprises, as it is precisely the balanced combination of monetary and nonmonetary resources which can guarantee both their autonomy of services and their economic viability.

Obviously, a combination also implies the reinvestment of earnings in the activity and the lasting collective ownership of these earnings, so that the profits made cannot be privately appropriated. This specificity explains why social enterprises adopt legal status like associations and cooperatives, presenting a nonprofit constraint or providing limits to the investors' profit. However, the social enterprises that present characteristics which have been a feature of the third sector for a very long time are original in two respects. Firstly, they form part of a new process of tertiarization of the economy, which is making their formation more complex than traditional third sector organizations. Groups of different categories of players (users, professionals, voluntary workers, etc.) are being formed around social enterprises, whereas associationism in the nineteenth century more often took the form of groups within one particular category (workers, consumers, or farmers). If their initial impulse comes from demand-side stakeholders, their development leads to a "new organizational form" which "may be called a *multi-stakeholders* or a multi-membership *organization* in order to emphasize that it is characterized principally by the composition of its social membership" (Borzaga and Mittone 1997, p. 13). Secondly, they maintain relations with the public authorities which are, by definition, different from those which characterized the period of spread of the welfare state. The concept of *hybridization* denotes not only the use of three types of resources, which is something that nonprofit organizations have been doing for a long time, but also the balancing of these resources in agreement with the partners and in accordance with the nature of the projects, whereas it used to be possible for socially useful work to be financed predominantly by redistribution.

Here, the rules governing exchange meet those governing production through the use of various forms of work. Whether it is done by users or activists, voluntary work does not replace employment creation. On the contrary, it can increase employment opportunities by lowering the price of services and helping to maintain a close and lasting link with users and other local partners. This use of voice rather than exit (Hirschman 1972; Pestoff 1994) is of great value in services to people because it makes it possible to lay down the quality criteria on the basis of a dialogue between providers, voluntary workers, and users (Pestoff 1994). The aim is to find the organizational arrangements which make it possible to deal with the problems of asymmetric information by maintaining this regular dialogue, which is considered to be the main source of information on the experience gained with the service.

What is revealed by the social enterprises which have gone far beyond the experimental stage is how the major problem of trust has led to an institutional innovation. This innovation is not merely the result of market mechanisms, but is based on co-construction which goes beyond the coproduction inherent in all services to take employment out of a domestic sphere and to enhance the standing of the tasks performed by employees. Moreover, it points the way to a limited rationalization which seeks to find the right balance between paid time and given time.

Social enterprises initially call into question the division in economic theory between *personal* and *collective services*, according to which personal services are services whose use is divisible (i.e., those in which users and use can be clearly identified), whereas collective services are indivisible because their use is "noncompetitive" (one person using the service does not impede others using the service as well) and "nonexclusive" (it is impossible or very costly to stop part of the population from gaining access to the service).

Some of the social enterprises are undoubtedly developing collective services in the traditional sense of the term, especially when they set up services intended to provide a solution to environmental problems by improving the management of natural resources.

Alongside these collective services, social enterprises also offer services that, although personal because their consumption can be divided, provide just as much collective benefit to the community. It is possible in this respect to speak of social usefulness or of *quasi-collective* services. The public regulation to which they are subject highlights the fact that their expected benefits do not just concern the private consumers who use them, because social justice criteria are involved or the external impact that they have on other operators has to be taken into account as well (Laville and Nyssens 1996).

Childminding is a typical example of services that go beyond the distinction between personal and collective. Childminding practices are very closely linked to the conception of the role and place of women in society. While these services have an impact on individual lifestyles, it is also true that the community influences the way in which apparently highly individual choices are made. Public authorities are thus entitled to intervene in the financing of childminding for reasons of equity that include the desire to make this service accessible to as many people as possible and to exercise control over the quality of services and because of the positive impact that these services may have on the community at large (for instance, increasing the availability of women in the labor market and the educational and preventive role played by childminding facilities). The same is true of other personal services.

Over and above the personal services that normally come to mind, other social enterprises may take on a quasi-collective dimension by providing services whose consumption is divisible, whether their objective is to offer a readily accessible cultural facility or to integrate disadvantaged people into a paid activity.

Even when services are personal, social enterprises consequently have a collective dimension. This is what gives them their overall consistency even though they are active in different fields. In overall terms, the dynamics that lead to their creation take two forms.

In the first dynamic, new services are set up *to meet an unsatisfied social demand*. Faced with a lack of appropriate answers to some problems from both private and public services, social enterprises have set up services as diverse as local community transport, cultural facilities, or day-to-day services.

Another dynamic that leads to the establishment of social enterprises is shaped by an initial desire *to integrate local populations and local areas*. For instance, development workers who are not local to districts can use their skills to locate the know-how which could lead to local economic development. The aim of these strategies is to bring about local development by locating and making use of local forces. In their concern to understand the actual situation of a region and to be accepted by residents so that these residents can be encouraged to play their part in economic projects, these strategies may require time. Comprehensive in-depth studies in several European countries (Fraisse et al. 2000; Laville and Gardin 1999) lead to the hypothesis of an ideal type of solidarity-based services to more precisely characterize these local initiatives and social enterprises.

9.2 The Ideal Type of Solidarity-Based Services

The initiatives which have just been mentioned differ from a commercially based approach, but have nevertheless proved their ability to establish relations based on trust between the parties in the services because of the guarantees they provide to both employees and users. If these initiatives are put into context, the result is the establishment of a hypothesis of ideal-type[3] "proximity services" in the context of a civil and solidarity-based economy, because economic initiative is founded on the desire to promote social relations based on solidarity.

This ideal type of solidarity-based services has two main characteristics:

- The services are designed through corresponding public spheres, which make it possible to shape supply and demand together.

[3]As expressed by Weber, who introduced this concept, "an ideal-type is obtained by emphasizing unilaterally one or more standpoints and by linking together numerous isolated phenomena ... arranged according to the previous, unilaterally chosen viewpoints in order to form a homogeneous framework of thought" Weber (1918, French translation, 1959, p. 180). This framework of thought is not an exact representation of reality, but emphasizes certain features for the purposes of research. The ideal type is not the same as reality, since it works out its constituent parts in order to define hypotheses more accurately and to characterize phenomena. In this way, the ideal type is a means of gaining knowledge, rather than an end in itself.

- Once the solidarity-based services have been established, they are consolidated by hybridization between the different types of resources involved, i.e., market, nonmarket, and nonmonetary resources.

9.2.1 Public Spheres for "Proximity Services" in Order to Permit the Joint Shaping of Supply and Demand

The idea of solidarity-based services implies breaking away from the situation in which those who require services are obliged to accept one of the formal services on offer or resort to moonlighting. Instead of each person attempting to privately solve, on an individual basis, the daily problems with which he/she is confronted, a determined effort is made to offer a solution by dealing with them collectively in the public sphere. In a gender perspective (Leira 1992; Lewis 1992), the fact that the civil and solidarity-based economy is located in the *public space* distinguishes it radically from the family-based economy. There is no intention of encouraging a return to the family through the civil and solidarity-based economy. The rural exodus or the increasing number of women in work show that the move away from a domestic economy has been a form of emancipation and reversing it is out of the question. Naturally, solidarity-based services are based on family resources, but they are designed to strengthen them rather than to confirm the isolation which can be faced by, say, women who care for elderly parents. For instance, the main objective of structures providing home help is therefore to preserve family stability. Professional assistance eases tensions by involving the elderly and their families in the preparation of a common project through discussions between these different stakeholders. The triangle formed by the association, the users, and the employees gives families an active role while at the same time encouraging more objectivity by having all those involved give thought to the question. As noted by Ben Ner and Van Hoomissen (1991), it is the position of users which is the key factor in determining which services are offered, be it on their own initiative, by their involvement with social entrepreneurs, or by the action of professionals who have become aware of unsatisfied demand through their experience with the provision of services.

Solidarity-based services have as starting point the daily practices of populations, the relations and symbolic exchanges which make up the daily fabric of local community life, and the aspirations, values, and desires of the people who use them. It is by considering these many aspects of reality in public forums for discussion (organized locally for this purpose) that supply and demand can be made to match. Service relevance in the eyes of the users can be explained by the fact that they go beyond the relationship with needs and requests as defined in market studies or needs surveys. Innovation in solidarity-based services is based on the use of a different economic principle from the market and redistribution, namely, the principle of reciprocity, which governs the process of interaction through which the services are organized. This *reciprocity* corresponds to the relationship between groups and individuals by means of services which derive their meaning from the desire to show a social link between the parties involved (Mauss 1950; Polanyi 1957, p. 19). As shown by Polanyi (1957), by identifying three economic principles to characterize the patterns in the relationship between economy and society (market, redistribution, and reciprocity), the reciprocal impulse is therefore different from market trading because it cannot be dissociated from the human aspect, which is tied up with the desire for recognition and power, and is distinct from redistribution because it is based on symmetry rather than centralization.

That is why solidarity-based services can, under certain circumstances, succeed where the market and the state have failed – they manage to reduce the asymmetric information problems between providers and seekers of services and to establish a relationship of trust with the user. Through the attention given to their form, these services can overcome the users' fears that their privacy will not be respected, thereby helping to formalize an extremely diverse demand while at the same time organizing supply. In this way, they can be said to jointly shape supply and demand.

9.2.2 A Hybrid of Various Resources

Not only do the solidarity-based services jointly shape supply and demand in public spheres for local services, but they also consolidate their position by combining monetary and nonmonetary resources.

Services which make use only of monetary resources, be they market or nonmarket monetary resources, have limitations in their activities when it comes to generating mutual confidence between users and providers in activities and commitment that include a dimension of intimacy. But nonmonetary resources, although they help the established structures to survive, will certainly not be enough to guarantee the future of these enterprises, because services which are too reliant on voluntary help tend to be nothing more than temporary phenomena dependent on favorable circumstances. For these reasons, hybridization is a way of consolidating services whose identity has already been asserted, as it is precisely the balanced combination of monetary and nonmonetary resources which can guarantee both the autonomy of services and their economic viability.

Obviously, such a combination also implies the reinvestment of earnings in the activity and the lasting collective ownership of these earnings so that the profits made cannot be privately appropriated. However, solidarity-based services that present characteristics which have been a feature of the third sector for a very long time are original in two respects. Firstly, they form part of a new process of tertiarization of the economy, which is making the formation of associations more complex. Groups of different categories of stakeholders (users, professionals, voluntary workers, etc.) are being formed around solidarity-based services, whereas associationism in the nineteenth century more often took the form of groups within one particular category (workers, consumers, or farmers). Secondly, they maintain relations with the public authorities which are, by definition, different from those which characterized the period of spread of the welfare state. The concept of hybridization denotes not only the use of three types of resources, which is something that associations have been doing for a long time, but also the balancing of these resources in agreement with the partners and in accordance with the nature of the projects, whereas it used to be possible for socially useful work to be financed predominantly by redistribution. This concept also expresses a combination between the market, nonmarket, and nonmonetary economies, which, by breaking down the barriers between them, internalizes the various externalities created by the services.

Here the rules governing exchange meet those governing production through the use of various forms of work. Whether it is done by users or activists, voluntary work does not replace employment creation. On the contrary, it can increase employment opportunities by lowering the price of services and helping to maintain a close and lasting link with users and other local partners. This use of voice rather than exit (Hirschman 1972) is of great value in services to people because it makes it possible to lay down the quality criteria on the basis of a dialogue between providers, voluntary workers, and users (Pestoff 1994). The aim is to find the organizational arrangements which make it possible to deal with the problems of asymmetric information by maintaining this regular dialogue, which is considered to be the main source of information on the experience gained with the service.

What is revealed by the ideal type of solidarity-based services, drawn up on the basis of real situations which have gone far beyond the experimental stage, is how the major problem of trust has led to an institutional innovation, which is not merely the result of market mechanisms, but is based on co-construction, which goes beyond the coproduction inherent in all services to take employment out of a domestic sphere and to enhance the standing of the tasks performed by employees. Moreover, it points the way to a limited rationalization which seeks to find the right balance between paid time and given time and between effectiveness and affection (Hochschild 1983).

The identification of the ideal type of solidarity-based services on the basis of practices shows the need for a socioeconomic analysis which does not merely mention the strong points of organizations resulting from their nondistributing profit constraint, but which takes a comprehensive look at the way in which they operate. The importance of the experiments referred to by this ideal type should therefore be evaluated simultaneously in three areas: social, economic, and political (Fig. 9.1).

9.3 Proposals for Reframing the Debate: Plural Approach and Theoretical Choices

This ideal type of solidarity-based services raises institutional questions which were not previously included in the nonprofit sector approach, limited to an organizational dimension. They refer to a plural economy and democracy, as well as welfare. They lead to theoretical choices, preferring the principles of solidarity and public action rather than the notion of sector.

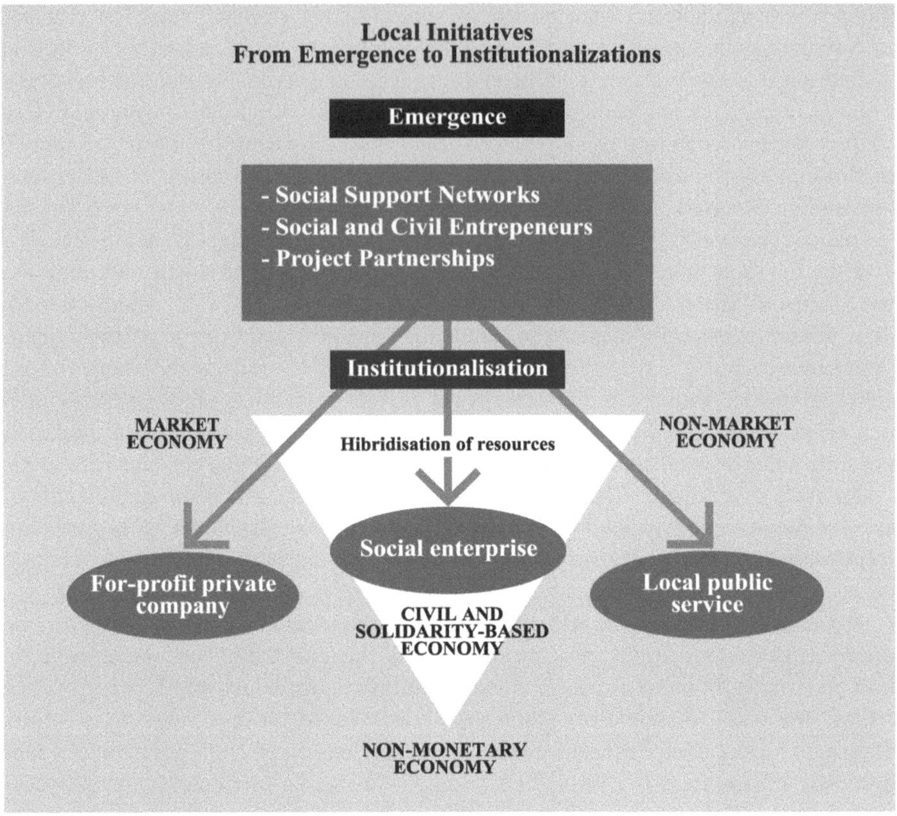

Fig. 9.1 Local initiatives. From emergence to institutionalization

9.3.1 The Socioeconomic Dimension

The synergism between state and market had derived from the possibility of job-rich economic expansion. But when growth could no longer ensure full employment, the welfare state was to undergo a deep crisis. As unemployment mounted, the state not only was deprived of part of its revenue but was called upon to commit funds to support the productive system, reduce joblessness, offer incentives for certain categories of the population to leave the labor market, and support vocational training. The dilemma in which the welfare state found itself magnified criticism of the principles that underpinned it.

What emerges from the analysis of strategies based on either free market or social democratic principles is that the debate is still obsessed with the respective roles of state and market. Trapped within these confines, it cannot lead to socially acceptable solutions. When the wage-earning society falters, proposals that rely on market mechanisms end up deepening social inequalities, while those that involve new roles for the state evoke fears of a latter-day managed economy under the guise of social service. New models can be found only by exploring the economy's hidden face, which is neglected by all the approaches that center on the state or the market.

To transcend the dilemmas arising from today's socioeconomic transformations, it is necessary to break loose from the conceptual framework in which the economy is identified exclusively with the principles of market forces and institutionalized redistribution as it was conceived in the nineteenth century. Indeed, this restrictive view masks the persistence in the modern economy of a third mode of exchange, *reciprocity*, as shown by Polanyi (1957). This third economic principle, different from the market and from redistribution, refers to the relationships established between groups or individuals via services that are meaningful only insofar as they reflect a determination to assert a social tie between the parties involved. They refer to a third component of economy, *a nonmonetary economy*. The social enterprise has to be placed inside this broader, tripartite economy to be fully understood.

The social enterprise achieves a mixture between the sense of local economy and the reference to more civic principles of equity and equality, what we in Latin European contexts express through the word *solidarity*. For this reason social enterprises can be considered as expressions of a *civil and solidarity-based economy*.

It is possible to say that the idea of a civil and solidarity-based economy is up to date because a large number of social enterprises have shifted the conventional boundaries between the economic and social spheres by limiting their scope to neither market economies nor state-sponsored solidarity. Despite their extreme diversity, all these social enterprises share common features with respect both to the parties involved and to their activities.

People are getting involved of their own free will to help carry out actions that contribute to the creation of economic activities and jobs while at the same time strengthening social cohesion. The entrepreneurial drive of the promoters of these schemes cannot be explained by expected returns on investment, but is based instead on a quest to forge new relationships of social responsibility through the activities that are carried out.

Such new economic activities that are successful show that they can thrive and flourish when they are supported by a *balanced combination of different types of resources* (market resources obtained through sales revenue, nonmarket resources derived from redistribution, and nonmonetary resources from voluntary contributions) and manage to establish complementarity between conventional employment and various forms of volunteer work.

The impact of social enterprises is therefore not confined to job creation. They also represent new interactions between the economic, social, and political spheres:

- Economically, they do not remain trapped by a "halfway house" conception whereby their sole function would be to facilitate reentry into mainstream employment for groups of people who have been shut out of it – they seek to broaden the economic domain via a wider range of action. While they can provide temporary jobs as a springboard for getting people back into work, they do not neglect permanent jobs as a means of entry into ordinary sectors of activity, and, above all, they create employment by starting up new activities.

Most importantly, however, employment is not treated as an end in itself, but only as part of a more comprehensive process which enables economic activity to be put back into a more meaningful context, in which the people employed have a genuine sense of belonging to a group. Services are produced not by hiring people in intermediate capacities or as household workers, in what are considered fill-in jobs, but by structuring activities in a collective framework which alone can guarantee not only the quality of services and jobs but the continued involvement of volunteers and customers alike. Rather than to defend employment at any price, whatever the pay and benefits to the employed, the idea is to institute complementarity between voluntary commitments and genuine jobs.

- Socially, schemes like these make it possible to nurture various forms of social networks via projects freely designed by their creators. Such solidarity gives impetus to networks whose growth is fostered by the increasing erosion of standards and values, withdrawal, and loss of identity. And yet networks like these do not signal a return to a kind of localism born of a denial of the social gains of modern times. On the contrary, they are joint undertakings whereby people interact to formulate solutions other than those offered by either market or state. They are rooted in a feeling of belonging; they seek to extract certain matters from the private domain and handle them in a public space with a view to remedying the inadequacies of both the private and public sectors.
- Lastly, on a political level, social enterprises get people directly involved in public affairs and help make democracy more vigorous because they are the result of ordinary citizens speaking out on the real problems they encounter. In addition, participants commit themselves to long-term relationships based on the freedom and equality of each member of the group, since all are encouraged to share their views and to get involved, regardless of status (as wage earner, volunteer, customer, etc.). Moreover, participants formulate projects that aim to achieve institutional change and not merely to produce; such schemes, because they constitute a "dimension of the public space in civil societies" (Evers 1993), emerge as new manifestations of democracy.

9.3.2 The Political Dimension

Civil and solidarity-based economy practices have begun to be identified in Europe as well as in America (Laville 1992, 1994). Challenging the institutional architecture, these practices reflect a quest for another relationship with the political sphere – one in which solidarity can be built through actions in which civil society and the state work hand in hand – and another relationship with the economic and social spheres, one in which economic initiative can pride itself on serving social aims as well.

Participants feel very strongly that democracy cannot be achieved solely by making a number of social corrections to market trends. For them, democratic relationships need to be promoted through economic initiatives. When market and state dynamics are not enough to create activities providing work for the active population as a whole, they need to be associated with complementary dynamics shaped by the citizens' reciprocal commitments so that the economic sphere becomes less selective.

In other words, one of the features of local initiatives is their desire *to further democracy at local level through economic activity*. This concern is reflected in a number of ways, for instance:

- By internalizing those social or environmental costs which are externalized by other enterprises. The goals of local initiatives mean that they take responsibility for functions such as the integration into employment of disadvantaged people and the long-term unemployed, as well as long-term development strategies for the maintenance of local heritage and the protection of the environment. In other words, they produce *positive collective externalities*.
- By respecting criteria of equity such as occupational *equality* between men and women or *accessibility* to the goods and services produced.

In this way, these initiatives are more than a simple reserve of jobs. They have a role to play in a development model that combines objectives of social cohesion and citizen participation since job creation within these initiatives is shaped by social and societal concerns.

The question that then arises is how to reshape government intervention in order to systematize support for economic activities that perform a social service and how to do it in a manner that reconciles initiative and solidarity. This means moving away from policies that "target" people and shifting toward project financing. This also means that it cannot be obtained by the good results of a few experiments – it requires sociopolitical pressure from civil society and needs to work with public bodies.

The contribution of the civil and solidarity-based services approach is indeed linked to the political dimension. In the nineteenth century, the extension of market-generated reactions on the part of society, among which the establishment of associations, was followed by the construction of a protecting welfare state. Associations were indeed "the first line of defense" (Lewis 1997, p. 166) elaborated by society before being relayed by the state.

One of the original features of the European point of view consists in integrating these initiatives of civil society into the public space of modern democratic societies. The relations between these initiatives and public authorities are then determining because they are linked to both political issues, the one that stresses the potential for action of the members of the political community as a whole and the one that is more centered on the exercise of power (Maheu 1991). Associations concern these two dimensions of politics – on the one hand, noninstitutional politics, centered on the potential for citizen action and which supposes that citizens make use in practice of the positive freedom to which they are formally entitled, and, on the other hand, institutional politics, centered on the exercise of power.

All the interactions between public authorities and civil society initiatives translate into mutual effects whose intensity and modalities greatly vary over time.

On the one hand, the initiatives of various social actors, through their very existence, contribute to the evolution of forms of public regulation. On the other hand, rules enacted by public authorities influence the trajectories of initiatives. Isolating organizations without grasping their relations with the public sphere amounts to forgetting their institutional dimension.

9.3.3 Theoretical Choices: Beyond a Sectorial Approach

The European approach replaces the reference to a sector by highlighting the structuring power of the principle of solidarity and by the study of the close relations between associative action and public authorities.

Contrasting with the hypothetico-deductive approaches which are characteristic of the neoclassical economy, a significant number of research works adopt more comprehensive methodologies. The main finding of these researches relates to a relativization of the notion of sector and to a mobilization of the concept of solidarity, to explain various social practices which can be grouped under the generic term of "civic associationism." Although they are not exclusively European, since quite similar orientations permeate the production of other continents, as evidenced by South-American literature (Larraechea and Nyssens 1994; Razeto Migliaro 1988; Ortiz Roca and Muñoz 1998; Cattani 2003), it is nevertheless worth noticing that European contributions to the analysis of associations are largely represented in this trend of research. In Europe, as a matter of fact, the recognition of human and civic rights destabilized the former social order without eliminating the differences of conditions inherited from traditional societies. With the apparition of the social question, as soon as the nineteenth century, the compatibility between citizenship and economic development consequently generated heated debates in the context of which associationist emergences occurred.

The second European originality consists in linking associationism and public action, since they have their roots in the resistance to the utopia of a market society and are deeply intertwined, whereas the diverse variants of theories of institutional choice considered associations as organizations intervening in case of market or state failures. A more historical analysis leads to highlight the fact that associative organizations "are not only producers of goods and services but important factors of political and social coordination" (Seibel 1990: 46). This is what the promoters of the international Johns Hopkins research project recognized when they passed from the notion of "nonprofit sector" to that of "civil society sector" (Salamon and Anheier 1997). This opening, which bears testimony to a rapprochement with the European view, is interesting as it accounts for the embedding of associations into society, but has, as its corollary, a too rapid assimilation of the associative sector with the whole civil society and does not sufficiently analyze the interactions with the state and the market.

As a matter of fact, a strictly sector-based vision fails to take into account the intermediary dimension of associations, which can be envisaged as spaces ensuring the passage from the private sphere to the public sphere. Associative action, born from the encounter among persons, opens up to the public space, i.e., it gives these persons the possibility to contribute to the construction of a common world which is necessary for democracy, through a voluntary commitment respecting the plurality of opinions, the conflictuality of interests, and the difference of perspectives. The mediation between private and public space, which happens in many different ways, and the mixture of resources and logics of action to which it refers, are poorly traduced by representations which suppose well-separated sectors, with clear-cut boundaries. The analysis of the genesis and institutionalization of associations underlines the scope of the interdependencies between associative action and public action.

If, following Cohen and Arato (1994), we define civil society as a sphere that is distinct from the state and the market, associations belong to an organized civil society, because they influence the configuration of the public space through innovations and dissensions that they manage to express in this public space, including their socioeconomic production. However, as rightly expressed by Barthélemy (2000, pp. 15–17), "the activities of the civil society cannot be dissociated from the political society," and associations are not only the expression of civil society, they are also implied in relations of power because they "publicize ideological conflicts of the global society, contribute to the formation of elites and to the structuration of local power and participate in the definition of public policies while legitimizing the political and administrative sphere."

Briefly stated, as Walzer (2000) notes, civil society, if it recognizes interpersonal links, is marked by inequalities (Chanial 2001). As far as the state is concerned – since it results from universalist orientations – it guarantees social rights while establishing general rules and standardized procedures which correct inequalities but also neglect the contribution of social relations of proximity. The real question does thus not concern the substitution of the state with civil society nor the dissolution of civil society in the market, but the mutual reinforcement between the democratization of civil society and the democratization of public institutions.

References

Barthelemy M (2000) Associations: un nouvel âge de la participation? Presses de Sciences Po, Paris

Baumol WJ (1987) Macroeconomics of unbalanced growth: the anatomy of urban crisis. Am Econ Rev 57:415

Ben Ner A, Van Hoomissen T (1991) Nonprofit organizations in the mixed economy. Ann Publ Coop Econ 62(4):519

Borzaga C, Mittone L (1997) The multi-stakeholders versus the non-profit organization. Universita degli Studi di Trento, Dipartimento di Economia, discussion paper, n° 7

Cattani AD (ed) (2003) Outra economia. Veraz Editores, Porto Alegre

Chanial P (2001) Justice, don et association: la délicate essence de la démocratie. La Découverte, Paris

Cohen JL, Arato A (1994) Civil society and political theory. The MIT Press, Cambridge, MA

Commission of the European Communities (1994) Growth, competitiveness, employment: the challenges and ways forward into the 21st century. Office for Official Publications of the European Communities, Luxembourg
Commission of the European Communities (1995) An investigation in the European Union. Local development and employment initiatives. Office for Official Publications of the European Communities, Luxembourg
Evers A (1993) The welfare mix approach – understanding the pluralism of welfare systems, report to the conference on well-being in Europe by strengthening the third sector, Barcelona, 27–29 May
Fraisse L, Gardin L, Laville JL (eds) (2000) Le fonctionnement socio¬économique du Troisième système. Recherche européenne pour la Direction de remploi et des affaires sociales (DGV) de la Commission des Communautés Européennes
Hirschman AO (1972) Voice (prise de parole) et exit (défection). In: Face au déclin des entreprises et des institutions. Les Editions Ouvrières, Paris
Hochschild A (1983) The managed heart, commercialization of human feeling. University of California Press, Berkeley
Larraechea I, Nyssens M (1994) L'économie solidaire, un autre regard sur l'économie populaire au Chili. In: Laville JL (ed) L'économie solidaire. Une perspective internationale. Desclée de Brouwer, Paris
Laville JL (1992) Les services de proximité en Europe. Syros, Paris
Laville JL (ed) (1994) L'économie solidaire. Une perspective internationale. Desclée de Brouwer, Paris
Laville JL, Gardin L (eds) (1999) Le iniziative locali in Europa: un bilancio economico e sociale. Bollati Boringhieri, Torino
Laville JL, Nyssens M (1996) Les services de proximité: un enjeu de société. Wallonie, Bruxelles
Leira A (1992) Welfare states and working mothers. The Scandinavian experience. Cambridge University Press, Cambridge
Lewis J (1992) Gender and the development of welfare regimes. J Eur Soc Pol 2(3):159
Lewis J (1997) Le secteur associatif dans l'économie mixte de la protection sociale. MIRE-Rencontres et Recherches avec la collaboration de la Fondation de France, Produire les solidarités – La part des associations, Paris
Maheu L (1991) Vers une grande théorie du politique. In: Maheu L, Sales A (eds) La recomposition du politique. L'Harmattan/Presses universitaires de Montréal, Paris/Montréal
Mauss M (1950) Essai sur le don, forme et raison de l'échange dans les sociétés archaïques. In: Sociologie et anthropologie. Presses Universitaires de France, Paris
Ortiz Roca H, Muñoz I (eds) (1998) Globalización de la solidaridad. Un reto para todos. GES/CEP, Lima
Pestoff A (1994) Beyond exit and voice, in social services. Citizens as coproducers. In: Six P, Vidal I (eds) Delivering welfare. Centre d'Iniciatives de l'Economia social, Barcelona
Polanyi K (1957) The great transformation. Beacon Press, Boston
Razeto Migliaro L (1988) Economia de solidaridad y mercado democratico, Libro tercero, Fundamentos de una teoria economica compensiva. Programa de Economia del Trabajo, Santiago du Chili
Roustang G (1987) L'emploi: un choix de société. Syros, Paris
Salamon LM, Anheier H (1997) The civil society sector: a new global force. Society 34(2):60
Seibel W (1990) Government/third sector relationships in a comparative perspective: the cases of France and West Germany. Voluntas 1:42
Walzer M (2000) Sauver la société civile? Mouvements 8
Weber M (1918) Essai sur la théorie de la science. French translation: (1959). Plon, Paris

Chapter 10
Socially Inclusive Urban Transformation After the Great Recession

Toward a New Civic Economy Model

Eugenio Leanza and Gianni Carbonaro

Abstract This chapter focuses on the need for a new civic economy model discussing first the causes and the conditions of the urban unbalances in Europe which are likely to persist in the aftermath of the Great Recession. It elaborates on the evolving role of cities within the EU and the mounting challenges for cities suffering from shrinkage and economic decline, considering the increasing aging of the European population. The authors argue that the effectiveness of centrally managed policy instruments to combat these trends is limited. So they make the case for the concept of a new civic economy enabled by a bottom-up, decentralized policy approach and sustained by innovative urban management practices, which they believe can address at least some of the challenges confronting European cities.

Keywords Great Recession • European cities • Public expenditure

10.1 Introduction

The effects of the Great Recession, climate change risks, the global shift in manufacturing activities toward Asia, and the rise of high-tech businesses driven by information technology, as well as rapid demographic aging, are some of the key challenges for transformational urban investment in the EU. The difficulties confronting the European continent have institutional, economic, and financial ramifications, but also global spatial impacts which are leading to a geographical

---

E. Leanza (✉)
Mandate Services Division, European Investment Bank, Luxembourg City, Luxembourg
e-mail: leanza@eib.org

G. Carbonaro
QASER Lab, University College London, London, UK
e-mail: gianni.carbonaro@gmail.com

G. Concilio, F. Rizzo (eds.), *Human Smart Cities*, Urban and Landscape Perspectives 19,
DOI 10.1007/978-3-319-33024-2_10

reconfiguration of the metropolitan and urban landscape in the EU (Sassen 2001, 2002; Glaeser 2011; Florida 2010; Leanza and Carbonaro 2013). Against this background, the need for a change in the urban investment paradigm should be more openly discussed among those interested in the future of European cities, who should become aware of the ongoing transformation of the European urban job markets, which in turn has serious consequences on the overall capital accumulation processes and the everyday life of EU citizens. It is time to focus on operational proposals for innovative solutions in cities, supporting the analysis through place-based approaches.

It is virtually impossible to describe in a short text the main challenges faced by the "shapers" of a participative approach to urban transformation in the EU's current situation, in which the unemployment rate is a double-digit number and most cities continue to suffer the effects of a persistently very weak economic situation. This takes place in the context of a broader economy which shows a tendency toward chronically deficient demand and job destruction in manufacturing/service sectors driven by technology advancement, which may point to a long-term trend labeled as "secular stagnation" by former US Treasury Secretary Larry Summers and other observers (Eichengreen 2015).

The current economic crisis forces us to review our understanding of how cities function and how urban growth and decline mechanisms have changed in the aftermath of the Great Recession. The answers to these questions can help us to think about how urban studies could be refocused in order to propose innovative ways to reinforce the job-creation function traditionally performed by urban areas in Western economies. Urban development is a key driver of economic development, and cities account for a major share (from 70 to 85 % in advanced economies) of what economists define as GDP. The overall urbanized population is expected to exceed the five billion mark, or 60 % of the world's total, in 2030. This compares with an urban population of 750 million, or less than 30 % of the world's inhabitants, in 1950 (Véron 2006), implying an overall exponential growth of urban economies, investment, and fixed assets, while urban land prices and financial wealth – mainly concentrated in thriving cities – expand at an even faster rate.

Increasingly and in the fast-growing, emerging market economies in particular, urbanization is characterized by the emergence of vast metropolitan areas. This tends to be justified by economists on the basis of (often difficult to demonstrate) agglomeration economies linked to the concentration of production factors and large and highly flexible urban labor markets, though in thriving economies small- and medium-sized urban areas still represent a vital and crucial component of the overall urban system (Glaeser 2009). In mature Western economies, urbanization processes generate a continuous, complex reshuffling in city hierarchies at the national and continental level, which are often driven by shifts in logistic and manufacturing capacity, while in emerging market economies urban growth is driven by rapid rural-urban migration (Saunders 2010; Davis 2006). These trends have fundamental macroeconomic, demographic, welfare, and environmental consequences and a long-term impact on the spatial concentration and dispersion of assets, including human and social capital. Urbanized surface has increased by 78 % in the EU since the mid-1950s, compared to a population growth of 33 %. Land taking and soil sealing "decoupled" from an overall view of sustainable urban land use are caused by ineffective planning practices and, possibly, by a lack of professional city management with a holistic understanding of the inner relations between the financialization of our economy, sustainability, resilience, efficiency, as well as urban demographics.

In this chapter, we will first discuss how the interlinked impacts of globalization, the financialization of the economy, the EU single market, and the currency union have contributed to the creation of serious urban unbalances in Europe, which are likely to persist in the aftermath of the Great Recession. Against this background, we then elaborate on the evolving role of cities within the EU, the mounting challenges for cities suffering from shrinkage and economic decline, and the effect of the wider aging of the European population. We argue that the effectiveness of centrally managed policy instruments to combat these trends is limited. Finally, we make the case for a bottom-up, decentralized policy approach sustained by innovative urban management practices, which we believe can address at least some of the challenges confronting European cities – the concept of the civic economy being one of the basic components of the new urban management paradigm.

10.2 European Cities After the Great Recession

10.2.1 Accelerating Urban Dualism in the European Union

The opportunities and challenges for the EU deriving from the progressive establishment and consolidation of a single market enabling capital, goods, people, and services to flow freely in the continent have been extensively discussed (Calafati 2014). However, only few in the professional and academic community have addressed the implications of the establishment of the single market on the competition among European cities and metropolitan areas. In a financial and globalized economy, in which barriers to the movement of labor and capital have been removed, cities are "read" by markets as competing functional urban areas, i.e., logistic, immaterial, human, and fixed capital platforms which underpin the production and exchange of goods and services in high-density job and energy markets. The effects of this "single market for cities" have been magnified by the introduction of a single currency, which has exposed the EU urban systems to shocks and adjustments driven by the wide differentials in total factor productivity between urban areas. Locational advantage within a unified market has affected the dynamics of a system originally based on the competition among firms located in different countries. In a sense, the dynamics of competition have been progressively evolving toward an "urban struggle" for global leadership and valuable resources among EU metropolitan systems, where the potential performance of locally based economic activities – integrating resources from the private, public, and third sector – affects firm and household location decisions and the future evolution of urban job markets.

This scenario is producing urban winners, in particular those cities which have proactively modified their strategies – attraction/retention of external trade, effective governance, and cooperative skills – in order to adapt to the new trends. However, competition among cities also leads to the unintentional emergence of a large number of unintentional urban casualties, with dramatic impacts on national finances. "Core" metro urban areas enjoy higher organic investment profitability, lower local systemic risks, direct investment flows, and the expectation of increasing or stable asset prices, which attracts further investment in a virtuous circle. Banks can soften lending rates to local projects in expanding urban economies, taking into consideration the lower default ratios and higher recovery rates. Surging land price dynamics can – through the appropriate use of value capture mechanisms – finance social expenditure and support infrastructure deployment without incurring additional public sector debt. In shrinking or declining areas, the opposite trends are at work, affecting long-term welfare levels.

A natural consequence of the EU institutional and monetary architecture – originally designed to work in a federal system – is that capital accelerates its movement from low-productivity toward high-productivity urban systems. In Europe the limited amount of EU budgetary funds allocated to social and territorial cohesion cannot significantly offset the self-reinforcing differentials in urban capital accumulation triggered by the wider macroeconomic developments. Thus, the burden of the adjustment falls upon the budgets of the national and local authorities and on their ability to design and implement structural reforms and innovative governance. In addition, urban crises often affect countries which are large net contributors to the EU budget, such as Germany, France, the UK, and Italy. This should encourage all urban actors to reflect on the need to look at the reorganization of EU urban space with a perspective which goes beyond a formalistic, region-focused, GDP-centered approach in the definition of European territorial cohesion priorities.[1]

The spatial reorganization phenomena just described do not differ in nature from those that took place in the USA already in the second half of the twentieth century, which led the economic profession to develop concepts such as spatial equilibrium and agglomeration economies to explain urban growth and decline in the American continent (for a recent synthesis, see Glaeser 2008, 2011). Under a single currency and a unified continental economy, the American urban geography and city hierarchy have been reshaped in many ways over the past 150 years. Agglomeration economies and changes in spatial equilibrium patterns linked to the creation of a single currency are similarly driving the current spatial adjustments in Europe at an accelerated pace.

[1]One of the most important authors pleading for the introduction of alternative approaches in the design of EU territorial cohesion policy is Laurent Davezies (2012). Harvey (2010, 2012, 2014) provides a radical vision of the reasons which have generated the spatial crisis.

10.2.2 Cities: Workshops, Playgrounds, or Betting Shops?

In recent years, there has been a lively debate among urban economists and urban geographers on whether the predominating role of the city should be seen mainly as a "workshop" – with urban economic development centered on the production side, which ultimately determines city dynamics, population changes, skill mix, and income levels – or as a "playground," where a key role is played by the presence of location-specific amenities which makes certain cities attractive to high value-added activities (Storper 2013). Moretti shows how the development of innovative sectors determines an acceleration of job growth in all market segments, with amenities playing a growing job-spurring role in thriving cities (Moretti 2012). The role of amenities has further benefited from the growing importance of financial transfers and accumulated financial wealth in advanced economies. Urban growth phenomena become partially disconnected from the direct productive role played by cities in the global exchange system of goods and services. In particular, the expansion of social welfare and pension systems, implying an enhanced consumption capacity and mobility among pensioners, has further reinforced the role of amenities in retaining and attracting people with spending power, leading to growing attention being paid to the provision of "customized housing" and "attractors" both within and outside productive hubs.

In a previous contribution, the authors have suggested that some of the recent wave of urbanization has been fueled through mechanisms better understood as a form of investment driven by debt-financed urban real options, closely connected to the financialization of modern economy (Leanza and Carbonaro 2015). Cities used to grow through relatively slow, continuous stratification, but in the twentieth century, urban investment became one of the key transmission channels of the central banks' monetary policy, both directly and indirectly through the effect created by financial and real wealth effects on household consumption (Mishkin 2001; 2007). In this context, access to urban building rights and the associated potential changes in land use in essence represent an entitlement to urban rent. Its (real option) value depends on expectations of future asset prices as well as monetary factors, including interest rate conditions, volatility and frequency of transactions, perceived locational advantages, asset demand/supply considerations, etc. In this perspective, investors – including homebuyers – target metropolitan growth poles with high volumes/density of transactions and sufficient asset liquidity, at the detriment of more peripheral cities, mainly in consideration of the high option value of urban investment in "core" growing cities. As a result, urban investment promoted by agents aiming to maximize the financial "up-side" potential via financially leveraged strategies has automatically determined an increased level of total debt and financial instability in urban economies. It should also be noted that "spatial" economic hedging via the creation of "city future markets" is intrinsically difficult, if not virtually impossible, due to the nature of real estate markets and their exposure to planning rights, which can be manipulated by the political authorities through the creation/allocation of new building rights "ex nihil" to vested interest groups. In this context, there are wide differences among urban areas in the elasticity of land markets to demand and supply factors. The growing importance of authorization processes in changes in land use (land, building rights, and gray costs in some emerging market economies can account for up to 50 % of the final building price) has also greatly strengthened the position of the supporters of liberalization and more transparent planning practices (Andersson and Moroni 2014). The control of concessions, building rights, or credit facilities can accelerate the extraction of financial extra profits favoring a limited number of individuals at the detriment of the remaining urban stakeholders. Investment decisions supported by debt-financed urban option models are particularly sensitive to fast growth/densification, price volatility, low interest rates, long exercise periods, and access to finance. However, a blockage in the powerful money machine of urban expansion triggers recessive processes which are often difficult to control. A still stumbling homebuilding industry has been, according to former Federal Reserve Governor Ben Bernanke, one of the main reasons why the US recovery has remained sluggish and failed to deliver job creation comparable to the traditional levels enjoyed in the American economy (Solow 2013).

10.2.3 Money and Cities

Money, credit, and monetary policy are not spatially neutral and have contributed to the acceleration of the urbanization rate in recent history. The monetary system operates as a powerful capital allocation mechanism by continuously transferring and concentrating economic and financial resources toward the most productive and efficient uses. However, when this mechanism is hampered by the decline in organic profitability, market economies react by moving development toward new areas, through spatial change and enhancements in technology and transport (Harvey 2014). The historical roots of the accelerated rate of urbanization at a global scale lie in these mechanisms, fueled by the progressive evolution of a dollar-led international monetary system and an increased financialization of the economy, nurtured by the large and persistent deficits of the US current accounts, the global currency adjustments after Bretton Woods, the recurring financial crises, and the current financial globalization. More recently, given the prevalence of accommodative monetary policies in most economies, the fate of many Western metropolitan areas has been exposed to a complex system of bets concerning the delocalization of production activities, the future evolution of agglomeration economies in different cities, and the consequent expected impact on urban asset and land prices.[2] In most cases urban planning and architecture become passive and

[2]The American 1993 Nobel Prize Laureate Robert J. Shiller has produced pioneering research in the field of the dynamics of asset prices.

oblivious followers of these trends, which are the ultimate drivers of urban growth and decline (Benevolo 2011; Campos Venuti 2010).

The loose monetary policies of most central banks and the broad availability of financial resources in search of speculative returns have pushed all urban players to modify their natural investment behaviors, favoring home ownership vs. renting, as well as stimulating an economic fight for the control of land destined to be urbanized/densified in consideration of its dynamic value potential. Even the social housing industry has started to transform its inner nature throughout Western economies, to take into account the economic agents' financial preferences for home ownership as a long-term saving instrument for households. These processes have produced a growing financial exposure to systematic "place-based" risks. While it is relatively easy to finance and realize urban expansion that is supported by credit expansion, which in turn is fueled by accommodative monetary policies, through the value capture mechanisms generated by the transformation of agricultural land into urbanized areas, it is more complex to obtain paybacks (net of depreciation and physical maintenance) on large-scale urban capital investment on a long-term, sustainable basis. This happens in the context of rapid aging processes, in which the financial holding preferences of middle-aged and retired people move from equity to fixed-income securities. Thus, in addition to the vision of cities as workshops or playgrounds, the "leveraged" betting shop may become a more appropriate image in certain cases (Minsky 1992).

10.2.4 The Mechanics of Urban Stagnation/Shrinkage

Typically, urban managers in shrinking or declining metropolitan areas are torn by conflicting needs such as implementing politically difficult cost cuts, reducing the burden of oversized and obsolete physical assets, urgently finding new adaptive cash-generating solutions. Identifying and promoting new roles for productive assets, including entrepreneurial resources and workers' skills, necessarily implies exploring opportunities in new directions – reusing underemployed assets, unlocking latent economic potential in neglected resources, repricing public assets and goods, eliminating redundant infrastructure, and supporting the civic economy through active human capital and employment policies. Wide-ranging and inclusive regeneration policies are likely to be needed for a successful metropolitan turnaround strategy, as stakeholders have to assume an increasingly proactive role. These efforts may also entail a significant rebalancing of the activities in individual cities and in the wider system of each city's spatial relationships. The ability to monitor and govern these changes and the capacity to benchmark and replicate success stories become a key to urban success.

Particularly in Europe, where banks play a dominating role as providers of finance, the main mechanisms which cause urban areas to experience a blockage in capital accumulation are through a squeeze in the banking provision of credit to local enterprises and projects. Technically this happens when the banks cannot achieve the targeted risk-adjusted financial return on their equity, on the basis of benchmarks defined by markets, shareholders' preferences, or political decisions (Modigliani and Miller 1958). Often credit crunches are also a result of a deficit in the banks' minimum regulatory capital caused by the accumulation of losses on their historical credit exposures. A deflationary situation is likely to have exacerbated impacts on economically weak cities, where decreases in the value or liquidity of assets held by banks as security (e.g., mortgaged houses) will affect the recovery rates in bankruptcy procedures (see Mian and Sufi 2014 for proposals concerning a reform of mortgage finance), leading to a contraction in the availability of local credit and an oversupply of assets, often triggered by unimaginative public repossession and foreclosure policies. Shrinkage in the economy can also be intensified by a fall in demand fueled by expectations of a deflationary fall in local asset and consumer prices, bringing down the number of jobs, salaries, and tax revenues while increasing the role played by high fixed costs in urban shrinkage. In these circumstances, there is very little that central banks can do in order to reactivate a virtuous economic cycle, and a substantial role is played by structural reforms which reactivate the potential of local economies. The long-term response – should it not be possible to activate the public sector's financial transfers – is a progressive rationalization in urban capital expenditure programs of both public and private sectors, as experienced in the American "rust belt" or Eastern Germany after 1989 (Katz and Bradley 2013; Tumber 2012; Coppola 2012; Ritter 2007).

These processes are rendered more acute by the consolidation of the banking industry into an oligopoly of large "global" players, partly as a result of bankruptcies, liquidations, or mergers of weak local banks. The globalization of the financial industry is accompanied by a progressive sector specialization, where asset management, pension and mutual fund activities, investment banking, credit origination, and payment services – segments characterized by healthier inner profitability and lower capital regulatory requirements – are progressively separated by "high capital-absorbing" credit activities (often externalized toward actors suffering from asymmetric information deficit). Against this backdrop, while well-functioning, integrated financial markets can be considered a positive feature in optimal currency areas, market-driven solutions do not factor in the suboptimal, fragmented nature of the EU currency union, which requires a strong presence of local banking and community impact finance, and may have destabilizing consequences, intensifying the unbalanced distribution of financial risks across EU territory.[3]

[3]See European Commission (2015).

10.2.5 Population Aging and Spatial (Dis)equilibrium in the EU

The challenges for urban turnaround strategies in declining European cities are further complicated by an aging population, which will affect their total factor productivity, labor market performance, organic cash generation capability, and fiscal balance. This is a central element in EU spatial reorganization, as the baby boomers represent today some 45 % of the labor force. This implies that, depending on the underlying demographic scenario, some 90–100 million individuals are going to retire over the next 20 years throughout Europe in a situation of high youth unemployment and weak public sector finance in most Member States.[4] The old-age dependency ratio is expected to double in the EU as a whole from four working-age individuals for every over 65-year-old to two, unless new immigrants bring about an increase in the workforce (as well as in expected labor revenue and tax generation) sufficient to compensate for the impacts of an aging indigenous population.

Aging will take place in a spatial context where new job creation will be unevenly distributed – also due to a lack of EU fiscal integration of social policies – depending on spatial variations in total factor productivity, innovation, and flexibility. In weaker urban contexts, actions aimed at stretching the working life of an aging population should be accompanied by investment in total factor productivity enhancing. Top-down actions aimed at alleviating the constraints posed by the increasing dependency ratio – e.g., through reforms reinforcing labor and women participation rates and the growth of microenterprise and small enterprises – should be accompanied by bottom-up "place-based" actions aimed at improving the effectiveness of the urban civic economy. Both approaches imply more investment in human capital and soft factors, rather than in hardware and fixed capital.

Population aging is already taking place and implies a downsizing in public sector expenditure due to the need to consolidate public accounts in more fragile economies in the Member States, in an environment where private sector job creation may also slow down due to lower demand in banking, insurance, finance, legal services, real estate, and construction sectors. Cities will be confronted with a reduction in the tax base and revenue generation capacity, i.e., factors essential for preserving, maintaining, and improving capital infrastructure, with a growing tension with central governments. Local labor markets will expand in cash-generating export sectors with a resilient productivity performance, but job growth should be accompanied and sustained by an enhanced performance of civic economy. It will therefore be essential to build a robust methodology to gauge total factor productivity trends and value creation in cities. Shrinking urban systems may end up exporting their human capital in the form of trained workers and specialized technicians, with the need to better assess the accumulation of human capital. New monitoring methodologies have to be implemented to assess the impact of brain

[4]For a recent report on this topic, see European Commission (2012). See also Sellon (2004).

drain. Some urban areas will be faced with an oversized, riskier, and decaying urban infrastructure. A lower total factor productivity will also bring about rising borrowing costs – a "local risk premium" charged on new initiatives to compensate for higher location-specific risks. In practice, in each metropolitan hub, the gap between investment profitability and the risk-adjusted cost of funds will have to be bridged and managed.

The demographic transformation will affect the costs of health services, social security, and the care for the disabled and the elderly, but also on the geographical distribution of financial transfers and pension payments. More resources devoted to these policies over the next 20 years may in some cases soften the territorial impact of aging for areas that are attractive retirement destinations (Bonvalet et al. 2007). The general lower capacity of national budgets to transfer resources from the center to the local economies is, however, an unwelcome development, considering that the impacts of economic crises generally tend to hurt cities more than the countryside. It is also to be noted that, in European nonfederal states, a substantial component of the public sector debt was generated by territorial transfers through the public sector's employment and welfare measures (Davezies 2012). Where overall population birth rates are stagnant or decreasing, accelerating migration flows and the ability to integrate new citizens become the main determinants of urban success or decline.

Spatial reorganization will lead to strongly dualistic wealth accumulation effects, since the new urban spatial equilibrium and the demographic transition are likely to destroy a substantial amount of wealth held by residents and investors in shrinking cities, while capital gains will concentrate in growing urban systems such as export hubs.[5] In the majority of EU countries, especially those of the so-called EU periphery, a major share of household wealth is held in urban assets, typically owner-occupied housing, as these have traditionally protected the value of the investment against inflation and economic downturns. In the case of Italy, for example, the housing component in total household wealth accounts for around 56 % in 2013, according to the Bank of Italy.[6]

Presently, and even more in the future, we are likely to experience an increasing importance of spatial policies. The reason is that, on the one hand, monetary policy does not appear to be working as well or as predictably as it did in the past. Quantitative easing seems to have only a limited impact on the long-term productivity of labor and capital, as the system appears superficially stabilized, but the conditions for a sustainable long-term allocation of capital are not met. On the other hand, fiscal policy could have a better and more long-lasting impact, but the EU – also because of self-imposed constraints – currently lacks the fiscal capacity to implement such policies. As a result of the limited effectiveness of traditional

[5]On this, see, for instance, McDearman and Liu (2012) and Dowell and Sung (2008).

[6]It is interesting to note that home ownership is more widespread in urban shrinkage contexts than in high growth poles (e.g., in the USA, Detroit has a home ownership in excess of 71 % vs. some 55 % in San Francisco).

monetary and fiscal policies, policies focused on managing spatial reorganization become critical in assisting economic adjustment in cities and through that in facilitating the overall recovery of the European economy.

As a result of the above-described trends, large numbers of young, talented professionals are increasingly attracted to high-productivity metropolitan hubs in the "core" economies, often located outside national boundaries. In a monetary union characterized by wide productivity differentials, firms must offer higher wages to retain high-quality human capital in disadvantaged cities and towns, thus deepening the competitive gap compared to metropolitan growth areas.[7] The emigration of talented workers leads to inflationary pressures on housing and *ceteris paribus* subdues salary growth in thriving metropolitan hubs. In the declining cities of the periphery, deflationary conditions and a likely fall in fertility prevail, due to the emigration of couples and women of childbearing age, as well as the pressures generated by accelerated aging processes.

Given the high level and expected dynamics of property prices, young households in growing metropolitan hubs have an incentive to "hedge" against the risk of further housing price increases by borrowing through long-term mortgages, thus taking on a costly and speculative financial burden on their future professional life income. Large personal borrowing, combined with long working hours and job competition, may lead young couples to defer the decision to have children to a later stage or to limit family size. Through this mechanism, turbulences in urban economic cycles have much wider and long-lasting impacts on national birth rates, population, and economic dynamics, which in turn, when not properly addressed, fuel populist rhetoric and conservative movements.

10.3 Innovations: The Challenges

10.3.1 Complementary Currencies

As we have argued in the previous sections, the pro-cyclical money supply coupled with deregulated financial markets and embedded speculative incentives (typical of our debt-based monetary economies) are not likely to be conducive to a stable and balanced urban development. Among some of the best-known reform proposals alternative to the current arrangements was the so-called Chicago Plan, proposed by Irving Fisher in the 1930s (in essence, moving from a banking-centered financial architecture toward a public sector monopoly) and, somewhat on the opposite side, Friedrich Hayek's monetary free markets proposed in his 1976 book *Denationalisation of Money* (Lietaer et al. 2012). As far as the euro is concerned, a number of experts have proposed to go back to a system of semi-floating currency rates within the EU, similar to the ECU Exchange Rate Mechanism in the 1980s and

[7] According to the assumptions of the spatial equilibrium theory.

1990s, to ease the restrictions imposed by the euro on the weaker economies, which under the current EU monetary and fiscal architecture can readjust only through long periods of deflation and austerity. Possibly one of the most relevant reform proposals in this sense has been advanced by W. Streeck, who considers that the euro could preserve an anchor role similar to the one originally proposed by Keynes for the Bancor during the initial Bretton Woods' negotiations (Streeck 2013).

In this context, it also appears of interest to consider the proposal by a number of unorthodox specialists to move toward "money ecosystems," characterized by the presence of parallel, complementary currencies alongside the conventional one.[8] This ecosystem would be characterized by a plurality of means of exchange and issuing institutions, including the government. According to the proponents of this alternative economic approach, regional complementary currencies would act as a medium of exchange, but not necessarily as a store of value (or unit of account) to mitigate the intrinsic spatial shortcomings of the euro monetary system.[9] These complementary currencies would operate according to the "demurrage" concept originally proposed by S. Gesell in Germany in the 1930s and experimentally tested by a number of small communities in Switzerland, Austria, and Germany.[10]

Some of the elements in these proposals are intrinsically disruptive and may involve radical changes in the urban capital accumulation cycle. For instance, large components of the debt-financed urban fabric – often one of the causes for the acceleration in CO_2 emissions – would become economically obsolete. However, in the current macroeconomic environment, the massive investments needed to develop alternative energy technologies necessary to cope with climate change will require complex financial incentives, which are likely to exceed the fiscal and borrowing capacity of most national economies. This may indicate that a further political evolution in our monetary arrangements is needed to cope with strategic sustainability issues. In the same way, some of the transformational challenges associated with the so-called smart city cannot be intended simply as the application of new technology to urban investment driven by traditional financial mechanisms, but involve a deeper rethinking of the financial and economic paradigms determining capitalistic urban development mechanisms. This is likely to imply a much lower prospective demand for urban assets in the form we currently know.

[8]There are various other explanations to justify the advantages produced by complementary currencies (Lietaer et al. 2012; Kennedy 1990). In our perspective, in a deflating open economy, the virtual disappearance of money from a single-currency monetary circuit can be assimilated to a sub-case of the Gresham Law applied to an open economic system. The lack of a means of exchange triggers the collapse of the debt-financed optional components of the urban system. The collapse of the Greek economy can be partly blamed on the blockage of the monetary activity in Athens, a city accounting for over half of the Greek economy's performance.

[9]See Munchau (2015).

[10]In essence, the application of negative interest rates for the possession of money, to avoid that the latter is hoarded by individuals, thus depressing consumption and investment. See Gesell (2003).

General purpose technologies, like those used in the "new machine age," are going to drive most of future economic growth through a cascade of complementary technological, economic, and organizational innovations. It is not certain that urban managers will be able to rapidly take advantage of smartness and flexibility, as proven by the research carried out by Brynjolfsson and McAfee (2014) on the introduction of electricity in the US economy, which took more than 30 years to start displaying its full effects. The newly introduced general purpose technologies based on computer science and artificial intelligence have four "game changer" dimensions – they are digital, exponential, combinatorial, and programmed for independent knowledge accumulation. It is fair to anticipate that the new technologies will find broad application in urban economies, but full gains in total factor productivity are likely to be captured only gradually.

As indicated by Larry Summers and others,[11] the main risks of the application of new technologies are connected to a productivity growth decoupled from job creation and wealth creation for the working class, as well as a risky disappearance of the middle class, which has traditionally represented the most active component of aggregated demand in urban contexts. After a phase of externalization of manufacturing jobs to low-wage countries, a large number of manufacturing/service jobs are being destroyed even in China – a destruction which is in large part due to routine tasks being taken over by increasingly competitive computer-driven machines and robots. These phenomena are replicated in software, services, media, finance, manufacturing, retail, trade, and healthcare – in essence, in all industries, affecting the "white collar" urban professional segments in particular.

In spite of the immense positive potential of technical innovation, the risks of technology-related disruptions are also with us, and we may have to radically rethink cities as living and working environments. A key challenge will be represented by the capacity of the institutions to preserve the integration between the work of specialists, the power of the new machines, and an equitable distribution of the new wealth produced. Scale will be less and less a condition for success, while the main objective may become to preserve the resilience of the job system and civic economy (Begg 2002). Technologies capable of generating a substantial number of innovative jobs will become crucial in this context. In cities, as in the wider economy, participative, liberal, and transparent institutions free from the influence of "extractive" vested interests seem to represent preconditions for a successful transition to a sustainable and inclusive long-term development (Acemoglu and Robinson 2012). In this context, the role of the civic economy warrants some reflection, which we will present in the next section.

[11] See, for instance, Summers (2013, 2014) and Summers and Balls (2015). See also Eichengreen (2015).

10.3.2 Civic Impact Finance and the New Urban Management

An innovative urban management approach should be based on two main elements: the first is the vision of the city – i.e., the relevant functional urban area (OECD 2013, 2009)[12] – as a set of interlinked assets. The second is the application of a "corporate finance" approach to the management of these assets, focusing on the current state of each asset, its associated cost and benefit streams, the aggregate asset inflows and outflows, and understanding their drivers, interdependence, and their impact on community behavior. The city diagnostics underpinning the definition of an investment strategy will, for instance, examine how the existing skill mix correlates with city demography, how aging may affect the city skill endowment in the medium to long term, and whether migrants will be able to replace the gap in younger population cohorts to preserve or improve the skill mix, to maintain competitiveness and the required cash generation capacity.

"Civic impact finance"[13] entails the interaction among "open data systems," financial instruments for the "participative smart city" which operate in those high impact investment areas and projects which cannot currently be served by the banking sector. The aim is to foster alternative, more affordable, and sustainable business models in sectors characterized by low financial returns. Examples can be found in impact finance, participative finance, and crowdfunding[14] models which could also target pension rights' notional capital (to be partly allocated following individual preferences) and "sustainable public tendering" by the public sector. A second pillar of civic impact finance is related to "holistic" urban sustainability reporting aimed at measuring the sustainable long-term impact of investments, with particular attention on those "soft" factors – like the accumulation and conservation of social and human capital – which are key determinants for a city's long-term competitiveness.

The utilization of urban complementary cryptocurrencies and the development of ICT-based payment systems and crowdfunding platforms at the metropolitan level may become some of the key tools to implement future smart city strategies and enable local civic economies to tackle urban decline. The utilization of

---

[12]Metropolitan functional areas are in essence highly integrated job systems. According to the authors, in urban economy, the performance of the working population (human capital) is deeply interconnected with four types of productive factors, namely, natural capital, fixed capital (housing, productive assets, logistics, and infrastructure), money and financial capital, and spatial capital (i.e., proximity and density factors which impact on the agglomeration economies and land rent). See Leanza and Carbonaro (2015).

[13]Of the many publications on civic finance, see Ahrensbach et al. (2011).

[14]Through the appropriate use of technology and innovation, the "participative smart technology" exercise may help to overcome some of the current problems inherent to what W. Streeck has defined as "Kaleckian conflict" between the "deployers" of capital/infrastructure (the "capitalists") and the "users" of capital/infrastructure (the "workers").

parallel currencies in metropolitan regions may also facilitate the deployment of environmentally friendly solutions which cannot meet the standard financial performance requirements of a market economy.

Although these innovations are conceptually appealing, implementing them in practice will require a change in the mindset of city managers. A shift will be needed from fragmented sector-specific decision-making to a strategic approach that takes into account an integrated view of asset management. This approach should be supported by a systematic analysis, in which territorial diagnostics (based on the vision of the city as a set of interlinked assets) is followed by the identification of priority investment, integrated plans for urban development, and project selection. This approach will shift long-term financing priorities from an emphasis on fixed assets to skills, human and organization capital, since return from the latter type of assets can outperform what can be achieved through traditional capital expenditure. Innovative city management is necessary to bring forth sustainable urban transformation in an increasingly risky, competitive, and volatile environment.

As an illustration, the integrated approach to sustainable modeling for urban asset management should encompass the transformation of the city energy system and the technologies involved. Currently the stimulus toward establishing decentralized energy systems and smart grids in Europe is primarily driven by energy producers, equipment suppliers, and utilities, usually without a focused analysis of the implications of smart city investing on city-level performance indicators, taking into account social affordability and economic externalities as well. Also to be noted, in view of the growing market integration generated by the globalization processes, the large metropolitan area governments will be confronted with the need to develop customized econometric models incorporating finance, land, and local job market performance to govern fiscal choices at an appropriate scale/timing.

10.3.3 The Role of Financial Instruments

The authorities responsible for urban transformation will have to improve their coordination with key urban stakeholders in order to establish the conditions for sustainable, nonspeculative growth. Ample opportunities exist to employ dedicated financial instruments more efficiently for smart, sustainable, and affordable cities. In this context, cooperation among authorities, the private sector, and local banks can lay the ground for a thriving civic economy.

Opportunities offered by EU financial instruments can start to progressively change the way EU budgetary resources are employed, encouraging the transition from traditional grant funding to revolving instruments capable of attracting additional financial resources and reconstituting the value of the invested capital, allowing for its further reuse. Tackling the European territorial dualism requires diversified investment strategies between the "core" and "noncore" cities, possibly even more than in a multicurrency macroeconomic setting, where devaluation could be relied on – and it cannot because of the single currency – to correct imbalances between countries.

In difficult areas the risk-weighted return of various types of urban projects cannot match the risk-adjusted cost of capital faced by banks and financial intermediaries. Here, EU financial instruments can be employed to facilitate the provision of finance at sub-commercial terms under EC-authorized state aid regimes. This will enable urban development funds and territorial financial instruments to offer terms and conditions which are compatible with the lower returns that characterize the environmentally friendly, job-creating, and socially inclusive projects of the civic economy. In challenging environments, financial instruments can be offered as subordinated mezzanine and junior capital, in order to modulate the financial risk profile to project performance and investor requirements. Such anchor projects with a tailored risk-reward balance are unlikely to be delivered through unassisted market mechanisms and could support the progressive shift of urban capital to socially productive uses, as well as lowering costs by selectively targeting efficiency gains.

10.4 Conclusions

Central banks have constantly used monetary policy with the aim of stimulating wealth, export rates, investment, and consumption, thus affecting how urban systems invest, typically acting on both the demand and the supply sides for housing, industrial equipment, and infrastructure. Keynesian "fine-tuning" investment, originally designed to stabilize the economic cycle in recessions, has progressively evolved into economic "push," pro-cyclical economic maneuvers. In the majority of market economies, economic growth has been pursued via planning-led "Keynesian-Fordian" urbanization, with large infrastructure projects often located at the edge of the cities, sometimes via the creation of satellite housing districts,[15] peripheral town network investment, and infrastructure serving wide-range metropolitan regions, in search of economies of scale, land rent, and agglomeration factors. While it is true that macroeconomic policies are not designed to affect territorial development, we have argued that monetary policy decisions are not spatially "neutral," as they redirect activity flows toward different territorial economies and project types and, with the increasing financialization of the economy, push city growth further toward debt-driven and speculative capital accumulation models.[16] In addition, in the wake of the Great Recession – especially in Europe – monetary policy no longer works as it used to. Wherever possible, central banks could be obliged to bypass a financial system (no longer able to complete its original capital allocation mission) and work directly with governments, to inject newly created money into the real economy through different channels involving new forms of monetary and banking

[15]Originally influenced by the Levittown social experiment in Nassau County (NY).

[16]On the risks associated to boisterous, debt-fueled urban growth in China, see Chen (2015).

innovation.[17] Even in these cases, the territorial implications of such interventions should be properly analyzed and openly discussed to ensure transparency and accountability and the potential impact on the democratic city models to be pursued with the active participation of entrepreneurs, workers, prosumers, and citizens.

Urban disciplines are confronted with various innovative approaches where concepts such as "sustainable civic transformation" (vs. "urban development"), "open systems" (vs. "closed systems"), "location-/place-based factors" (vs. "sector-centered/horizontal policies"), "tailored impact investing" (vs. "debt-financed infrastructure"), "nonlinear/disequilibrium" (vs. "linear/equilibrium"), and "combinatorial economies" (vs. "scale economies") are gaining importance and underpin the future transformation of urban areas and metropolitan systems in advanced economies. In this context, urban management and city system modeling, planning, and architecture need to be increasingly interlinked with other disciplines in order to address issues related to sustainability, complex networks, value systems, social behavior, institution building, and the need to reinforce human and social capital in cities.[18] Educational and training institutions should develop these innovative approaches in order to enable professionals to work in a new multidisciplinary environment.

We are confronted with a transformation of the leadership in our economies, from natural resource owners of the past, to urban and financial asset owners of the present, to technology and intangible factor (represented by the digital economy) owners of the coming age.[19] The traditional urban transformation model, controlled by the public sector and based on Keynesian-Fordian infrastructure stimulus, seems to be broken, and current banking and financial instruments do not seem able to provide an appropriate solution. What comes next, then? Cities built according to the mass production debt-fueled paradigm must develop new, place-based, customized transformation strategies aimed at achieving a more balanced management and evolution of their multiple, integrated assets which represent the main source of their citizens' wealth and earning capacity in the future. The links between efficiency, resilience, diversity, and connectivity in cities and their impact on economic inequality and social strains are not explored enough in the professional and political debate. The impact of accelerated urban change and the challenges it will pose to future generations are probably underestimated, one simple reason for this being that perhaps there is no clarity on how to capture or measure this impact – and something we do not perceive or understand is often simply ignored. An open discussion is

[17]See Chick V., Graeber D., Wren-Lewis S., and other 16 economists (2015) "Better ways to boost Eurozone economy and employment," Financial Times, 26 March.

[18]In an urban context, social capital could be regarded as the ability of local/urban communities to preserve certain social bonding qualities, behaviors, or performances despite change in demographic fundamentals. On the definition of social capital, see Putnam (1995) and Stiglitz (2012) on the socially and economically disruptive consequences of inequality, while the classical insights of Jacobs (1961) on the role of local communities and neighborhoods in producing a vibrant city economy remain as authoritative today as they were when they were originally written.

[19]Rampini (2014).

difficult without appropriate knowledge, data, models, and institutions, making it hard for urban managers, political leaders, and financiers to make correct choices in their strategies.

Therefore there is a need for a new urban management paradigm which incorporates the concept of civic economy and reconciles the financial imperatives of complex advanced economies with long-term sustainable development. Perhaps new or reformed institutions are needed to direct the transformation of our cities so that, for instance, growing wealth inequality between and within cities does not bring down the overall quality of life in our society. The key to move toward an effective civic economy, capable of succeeding where top-down centralized policies seem to have failed, may lie in engaging more in microscale optimization processes and experiments, based on the development of interdisciplinary, location-specific know-how, benchmarked against the best practice emerging in peer communities, and benefitting from the close interaction offered by institutional innovation, social media, and other emerging ICT innovations.

Acknowledgements The authors gratefully acknowledge the decisive contribution of Piotr Pomorski, whose assistance in shaping this paper was essential. The responsibility for the views and opinions contained in this paper rests solely with the authors, and such views and opinions do not necessarily represent the position of the European Investment Bank.

References

Acemoglu D, Robinson JA (2012) Why nations fail: the origins of power, prosperity, and poverty. Crown Business, New York

Ahrensbach T, Beunderman J, Fung A, Johar I, Steiner J (2011) Compendium for the civic economy: what the big society should learn from 25 trailblazers. 00:/in Association with NESTA & Design Council CABE, London

Andersson DA, Moroni S (2014) Cities and private planning. Property rights, entrepreneurship and transaction costs. Edward Elgar Publishing Limited, Cheltenham

Begg I (2002) Urban competitiveness: policies for dynamic cities. Policy Press, Bristol

Benevolo L (2011) La fine della città. Laterza, Bari

Bonvalet C, Drosso F, Benguigui F, Huynh P-M (2007) Vieillissement de la population et logement: Les strategies résidentielles et patrimoniales. La Documentation Francaise, Paris

Brynjolfsson E, McAfee A (2014) The second machine age. Work, progress, and prosperity in a time of brilliant technologies. W. W. Norton & Company, New York

Calafati A (2014) La città nel progetto europeo. In: Della Torre M, Pedretti B (eds) Cittadinanza. Geografie, filosofie, iconografie, economie. Donzelli editore, Roma

Campos Venuti G (2010) Città senza cultura. Laterza, Bari

Chen Z (2015) China's dangerous Debt. Foreign Aff. 94(3):13–18

Coppola A (2012) Apocalypse Town. Cronache dalla fine della civiltà urbana. Gius, Laterza & Figli, Bari

Davezies L (2012) La crise qui vient: La nouvelle fracture territoriale. Seuil, Paris

Davis M (2006) Planet of slums. Verso, New York

Dowell M, Sung Ho R (2008) Aging baby boomers and the generational housing bubble. J Am Plann Assoc 74(1):17–33

Eichengreen B (2015) Secular stagnation: the long view. Working paper 20836. NBER, Cambridge, MA

European Commission (2015) Green paper – building a capital markets union, COM 2015-63 final. European Commission, Brussels
European Commission Directorate-General for Economic and Financial Affairs (2012) The 2012 ageing report. Economic and budgetary projections for the 27 EU member states (2010–2060). European Commission, Brussels
Florida R (2010) The great reset: how new ways of living and working drive post-crash prosperity. HarperCollins, New York
Gesell S (2003) Die natürliche Wirtschaftsordnung durch Freiland und Freigeld. Cologne (digital edition)
Glaeser E (2008) Cities, agglomeration and spatial equilibrium. Oxford University Press, Oxford
Glaeser E (2009) Why has globalization led to bigger cities? The New York Times, 19 May
Glaeser E (2011) The triumph of the city: how our greatest invention makes us richer, smarter, greener, healthier and happier. Penguin Group, New York
Harvey D (2010) The enigma of capital and the crises of capitalism. Oxford University Press, Oxford
Harvey D (2012) Rebel cities: from the right to the city to the urban revolution. Verso, New York
Harvey D (2014) Seventeen contradictions and the end of capitalism. Profile Books Ltd, London
Jacobs J (1961) The death and life of great American cities. Vintage Books, New York
Katz B, Bradley J (2013) The metropolitan revolution. How cities and metros are fixing our broken politics and fragile economy. Brookings Institution Press, Washington, DC
Kennedy M (1990) Geld ohne Zinsen und Inflation. Verlagsgruppe Random House, München
Leanza E, Carbonaro G (2013) Making European cities more affordable, productive and sustainable. L'industria 2:275–294
Leanza E, Carbonaro G (2015) Attaining sustainable, smart investment. The smart city as a place-based capital allocation instrument. In: Vesco A, Ferrero F (eds) Handbook of research on social, economic, and environmental sustainability in the development of smart cities. IGI Global, Hershey (in press)
Lietaer B, Arnsperger C, Goerner S, Brunnhuber S (2012) Money and sustainability. The missing link. Triarchy Press, Axminster
McDearman B, Liu A (2012) Ten steps to delivering a successful metro export plan. Brookings Institution-Rockefeller Foundation, Washington, DC
Mian A, Sufi A (2014) House of debt: how they (and you) caused the great recession, and how we can prevent it from happening again. University of Chicago Press, Chicago, IL
Minsky HP (1992) The financial instability hypothesis. Working paper (74). The Jerome Levy Economics Institute of Bard College
Mishkin F (2001) The transmission mechanism and the role of asset prices in monetary policy. The National Bureau of Economic Research, Cambridge, MA
Mishkin F (2007) Housing and the monetary transmission mechanism. The National Bureau of Economic Research, Cambridge, MA
Modigliani F, Miller MH (1958) The cost of capital, corporation finance and the theory of investment. Am Econ Rev 48(3):261–298
Moretti E (2012) The new geography of jobs. Houghton Mifflin Harcourt, New York
MunchauW (2015) Why smoke and mirrors are safer than cold Turkey. Financial Times, 16 March
OECD (2009) How regions grow: trends and analysis. OECD, Paris, France
OECD (2013) Definition of Functional Urban Areas (FUA) for the OECD metropolitan database. OECD, Paris, France
Putnam R (1995) Bowling alone. America's declining social capital. J Democr 6(1):65–78
Rampini F (2014) Rete Padrona, Amazon, Apple, Google & co. Il volto oscuro della rivoluzione digitale. Giangiacomo Feltrinelli Editore, Milano
Ritter GA (2007) Der Preis der deutschen Einheit. Die Wiedervereinigung und die Krise des Sozialstaats. Verlag C. H. Beck oHG, Munich
Sassen S (2001) The global city: New York, London, Tokyo. Princeton University Press, London
Sassen S (2002) Global networks, linked cities. Psychology Press, London

Saunders D (2010) Arrival city: the final migration and our next world. William Heinemann, London
Sellon GH (ed) (2004) Global demographic change: economic impacts and policy challenges. The Federal Reserve Bank of Kansas City, Kansas City
Solow R (2013) How to save American finance from itself. Has financialization gone too far? New Republic, 08/04/2013
Stiglitz JE (2012) The price of inequality: how today's divided society endangers our future. W.W. Norton & Company, New York
Storper M (2013) Keys to the city. How economics, institutions, social interaction, and politics shape development. Princeton University Press, Princeton
Streeck W (2013) Gekaufte Zeit. Die vertagte Krise des demokratischen Kapitalismus. Suhrkamp Verlag, Berlin
Summers LH (2013) Economic possibilities for our children. NBER Reporter (4), 1–6. The National Bureau of Economic Research, Cambridge, MA
Summers LH (2014) Reflections on the new 'Secular Stagnation hypothesis'. Vox, 30 October.
Summers L, Balls E (eds) (2015) Report of the commission on inclusive prosperity. Center for American Progress, Washington, DC
Tumber C (2012) Small, gritty and green. The promise of America's smaller industrial cities in a low-carbon world. MIT Press, Cambridge, MA
Véron J (2006) L'urbanisation du monde. La Découverte, Paris

Part IV
Experiences

Chapter 11
The Human Smart Cities Manifesto: A Global Perspective

Àlvaro Duarte de Oliveira

Abstract Cities are progressively adopting information and communication technologies (ICTs) to ensure that their critical infrastructures and utilities are managed more efficiently, thus becoming smart cities. In this sense, the concept of smart city has created a new market opportunity for the traditional ICT industry, focused on the physical and technical endowments and neglecting the truism that cities are made of people. Not surprisingly, 'technology-pushed' solutions have often failed to engage the citizens and the public authorities themselves, who didn't take ownership of the 'smart' services experimented in this way. A claim for democracy, innovation and participation is becoming increasingly pressing, establishing the need to 'listen and talk to the streets' and ultimately changing the governance paradigm. These challenges call for a transformation in the way we all work, live, play and build our future, which in turn places a special burden on those holding the responsibility to govern such processes with an optimum usage of the public resources available. The reality is therefore that cities are only smart when they manage to take full advantage of the human capital of their citizens, creating innovation ecosystems where the new dynamics of wealth and job creation takes place and promotes new forms of participatory governance, in short, when they become Human Smart Cities. The Human Smart Cities Manifesto, launched during the Forum for Public Administrations in Rome, June 2013, is presented.

Keywords (Human) smart cities • ICTs • Human capital

11.1 Context

Cities face new challenges every day to create prosperity and ensure good quality of life to their citizens in a world increasingly adopting advanced communication infrastructures and technologies. The emergence of the digital society creates a

---

À.D. de Oliveira (✉)
Alfamicro, Sistemas de Computadores lda, Caiscais, Portugal
e-mail: alvaro.oliveira@alfamicro.pt

G. Concilio, F. Rizzo (eds.), *Human Smart Cities*, Urban and Landscape Perspectives 19,
DOI 10.1007/978-3-319-33024-2_11

tension between threats to established social cohesiveness and ways/opportunities of living in community and building sustainable economies. As cities grow, the uncertainty about the social models resulting from the digitalization of the society calls for a particularly decisive participatory action from public and private city authorities. This uncertainty is exacerbated by megatrend of megacities, with the world human population living in cities going from 50 to 70 % by 2050. In such scenario, the well-being and quality of life of citizens are impacted by challenges such as demographic shifts, gentrification, sustainable housing, mobility, environmental impact, food and water sustainability, health-care support and security and safety. Solutions for these challenges became a priority.

Cities are also progressively adopting information and communication technologies (ICTs) to ensure that their critical infrastructures and utilities are managed more efficiently, thus becoming smart cities. The citywide use of sophisticated ICT infrastructures capable of sensing what is happening in a car park, a traffic jam, available hospital beds, energy dissipation, water or air quality, temperature, noise, etc., coupled with models and data analytics, processed most likely in computing clouds, completes the emerging picture of the 'city as a machine' and allows for acting in the real world so as to adapt it to new circumstances. As a result of these actions, cars can be directed to the available parking places and avoid congested zones, ambulances can be readdressed, unnecessary consumption of energy can be rationalised, citizens can be warned of the changes in environmental conditions, etc.

In this sense, the concept of smart city has created a new market opportunity for the traditional ICT industry, focused on the physical and technical endowments and neglecting the truism that cities are made of people. Not surprisingly, 'technology-pushed' solutions have often failed to engage the citizens and the public authorities themselves, who didn't take ownership of the 'smart' services experimented in this way.

More generally, urban challenges are bigger and call for a more radical transformation than what can be achieved by technology alone. Examples of 'wicked problems' come from areas of interest (and legal competence) of local governments, such as waste recycling, water and energy savings, collective mobility, public safety and health and social care. A claim for democracy, innovation and participation is becoming increasingly pressing, establishing the need to 'listen and talk to the streets' and ultimately changing the governance paradigm. These challenges call for a transformation in the way we all work, live, play and build our future, which in turn places a special burden on those holding the responsibility to govern such processes with an optimum usage of the public resources available. The reality is therefore that cities are only smart when they manage to take full advantage of the human capital of their citizens, creating innovation ecosystems where the new dynamics of wealth and job creation takes place and promotes new forms of participatory governance, in short, when they become Human Smart Cities.

11.2 Concept

An illustrative and concrete example of the Human Smart Cities approach has been developed in the *Periphèria* project (2012), which has been published online and in a printed book. This methodology supports the process of citizen engagement and gives motivations for collaboration in the codesign and co-creation of civic solutions.

It is important to point out that the implementation of the Human Smart Cities concept can be made through the use of frugal technology and does not always require sophisticated and complex infrastructures. This fact is relevant essentially in what concerns the scalability of the solution. Simple and creative solutions can emerge from the local communities, which allow, as an example, big cities to extend their strategies and include broad metropolitan areas or small cities to integrate new strategies. This is an important advantage for the city administration that has the potential to enable the creation of humanly smart services without having to make significant investments. Another significant advantage of this concept, from the governance point of view, is the fact that the codesign and coproduction of solutions take out the 'burden' of the city administration processes that become lighter and more transparent.

Figure 11.1 shows the Human Smart Cities ecosystem of urban innovation (also called urban living lab), which applies user-driven open innovation methodologies and tools to the codesign and coproduction of social and technological innovation services and processes, by citizens and governments together, in order to solve real problems of common concern.

A service platform run by the city administration promotes the formation of communities and the collaboration amongst them. These communities are initially virtual, but then soon encouraged to migrate to the physical environment to meet together and discover their common wishes, interests and needs. The engagement of citizens in idea generation is essential to build a trusted environment in which the community and government codesign solutions together. Most of the time, the city administration only gets feedback from a small number of citizens; thus, it is important to put in place strategies to 'listen and talk' to all the groups of citizens.

11.3 The Human Smart Cities Manifesto

The Human Smart Cities approach is gaining increasing support from city governments across Europe and beyond as well as the smart city research community, as it more effectively addresses key needs related to low-carbon strategies, urban environment, sustainable mobility and social inclusion through a more balanced, holistic approach to technology and social innovation.

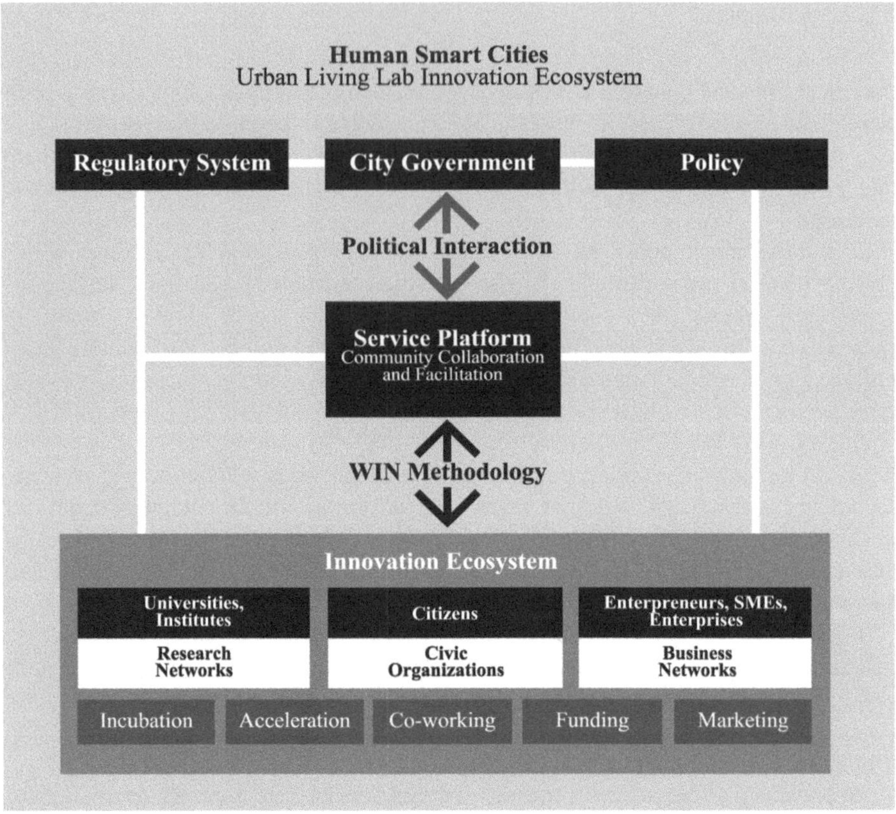

Fig. 11.1 Urban living lab innovation ecosystem

In order to promote the exchange of experiences and adopt similar practices, a Human Smart Cities Manifesto was publicly announced and signed in Rome on 30 May 2013, during the Forum PA Conference. With this Manifesto, motivated by the main challenges that cities, all around the world, are facing today, the signatories agree to adopt the UN standards of governance, promoting participation, decency, transparency, accountability, fairness, efficiency and sustainable development. The text of the Manifesto was as follows:

Preamble We, the signatories of this Manifesto, come together to address the three main challenges facing the cities around the world today:

- The devastating effects of the financial crisis undermining the European social model. This is leading to severe limitations in cities' abilities to invest in new infrastructures, and in some areas even for the provision of basic city services such as transportation and social services.
- The increasing threat and disruption brought about by climate change to our territories. As major floods and droughts become ever more common, the environmental effects of urbanisation and the lack of adequate tools and behaviour patterns becomes increasingly evident.

- The demand for more effective representation set forth by our constituencies. The so-called democratic deficit is a cause for alarm for governance at any scale, but it also adds to the difficulty of building trust and engaging citizens in addressing common problems.

These challenges call for a transformational change in the way we all work, live, play, and build our future, which in turn places a special burden on those of us holding the responsibility to govern such processes with an optimum usage of the public resources available. We are deeply convinced that technological and social innovation can make an invaluable contribution in that direction, if urban policies adequately consider citizens and their innovation capacity the most valuable resource.

In this crucial time and with these challenges in mind, we reach out to our citizens and enterprises to join us in a broad endeavour of co-creating the most appropriate strategies for each of our cities, as well as implementing them jointly in the years to come.

Vision Human Smart Cities are those where governments engage citizens by being open to be engaged by citizens, supporting the co-design of technical and social innovation processes through a peer-to-peer relationship based on reciprocal trust and collaboration. The Human Smart City is a city where people – citizens and communities – are the main actors of urban "smartness". A Human Smart City adopts services that are born from people's real needs and have been co-designed through interactive, dialogic, and collaborative processes.

In a Human Smart City, people are not obliged to adopt technologies that have been selected and purchased by their municipal governments; they rather are encouraged to compose their own services using available technologies in simple, often frugal solutions. Co-creation initiatives at the heart of the Human Smart City concept also stimulate local development, creating new business models and new apps, products, services and solutions. Indeed, the solutions for the big challenges of our time require not only innovative technologies but, above all, mass behaviour transformation of the kind that can only be achieved through the involvement of people. Through the appropriate governance of social and technical innovation and the integration of Future Internet technologies, Living Labs and Social Innovation, the Human Smart Cities vision aims to build on a new sense of belonging and identity, wellbeing and community, to shape a better and happier society.

Partnership With this **Human Smart Cities Manifesto**, we signatories join forces to build a network of Human Smart Cities – the HSC Network – throughout the world that share this common vision and learn from each other to find the right path towards social and urban innovation. In accordance with the Human Smart Cities Roadmap, we agree to the following 7 commitments.

1. We will adhere to the key values of the HSC network – trust, openness and transparency, enabling networking and participation; human-centricity; bottom-up, enriched communication; co-creation and collaboration – in all our activities. All participant cities are considered to have a unique role, contributing to innovation initiatives and policies on an equal, peer basis.
2. We will apply the Open Government model to our city's use of ICT, including transparency and Open Data, an appropriate role for Open Source and re-usability, and citizen and stakeholder participation in decisions related to key ICT infrastructures and services. Where possible, we will favour the adoption of simple, frugal solutions that can be shared across the HSC network.
3. We will explore where possible the citizen-centred approach for the co-design of all new city services, promoting creativity and engagement as well as active participation in service delivery. Together with our economic partners, we will also explore potential new business models for the promotion of innovation ecosystems and the delivery of services in the public interest.

4. We will promote institutional innovation within and across our city administration as an integral part of our role in service co-design. This includes the exploration of Pre-Commercial Procurement and other procedural and financial innovations, and collaboration with regional and national authorities to promote policy coherence across instruments and programmes.
5. We will actively participate in networking among signatories to this manifesto, actively contributing to its shared resources, attending the HSC network's yearly conferences, and collaborating to define a sustainable institutional structure. We will also leverage the potential of relevant networks at both international and national level.
6. We will together define measurable goals, success criteria, and performance indicators allowing our stakeholders to assess their progress towards objectives and promote scaling up and transfer. In sharing our evaluations, we will aim to promote learning from different cultural and urban contexts rather than competition, while still demonstrating the concrete benefits of the HSC approach.
7. Finally, we will promote the HSC network itself globally, as an innovative and open multi-level partnership ideally suited to implementing bottom up the policy goals of Europe 2020 and similar frameworks. To that end, we commit to bringing one new signatory per year.

Launched in Rome, 29 May 2013, by the partners of the Periphèria EU Project

The Manifesto initiative led in October 2013, in Bologna, to form the Human Smart Cities network, which was initially launched having 70 cities expressing their interest in membership. The main goals of the Human Smart Cities network are to allow cities to learn from each other, accelerating the process of social and urban innovation; to promote knowledge sharing between cities; to promote regular events so that knowledge sharing is potentiated; and to create a dynamic platform that includes a database of case studies.

Chapter 12
Sletten: Rethinking Urban Habitats Through Creative Management and Social Engagement

Stefan Darlan Boris and Peter Gall Krogh

Abstract Landscape laboratory is a concept for citizen involvement in the maintenance and development of urban forests. The concept has been developed and refined since the mid-1980s and now counts for four such places in the Scandinavian countries. It is widely recognized that the yet not peaked urbanization poses challenges for natural habitats in urban areas; landscape laboratories as urban forests are a response to that.

The chapter reports on the experiences of establishing an urban forest as a landscape laboratory simultaneously to developing a housing project, called Sletten situated in Holstebro, Denmark. The chapter brings forward three emblematic exemplars of how the place is appropriated by the people living there. Sletten is both an actual suggestion for novel neighborhood design and a full-scale experimental platform where scientists and practitioners with different or no professional backgrounds can meet and collaborate on testing and developing new concepts for the design, establishment, and management of urban forests.

Keywords Urban forest • Landscape laboratory • Stakeholder engagement • Community resilience • Decentralized responsibility

12.1 The Context

The Danish municipality of Holstebro established Sletten, a combined suburb and urban forest, in 1996. In addition to eight small, village-like clusters of houses, Sletten is part of a network of landscape laboratories in the Nordic countries

S.D. Boris (✉)
Platform Urbanism and Landscape, Aarhus School of Architecture, Nørreport 20, 8000 Aarhus C, Denmark
e-mail: stefandarlan.boris@aarch.dk; sdb@aarch.dk

P.G. Krogh
Department of Engineering, Head of Design Research, Navitas Inge Lehmanns Gade 10, 8000 Aarhus C, Denmark
e-mail: pkrogh@eng.au.dk; www.eng.au.dk

G. Concilio, F. Rizzo (eds.), *Human Smart Cities*, Urban and Landscape Perspectives 19,
DOI 10.1007/978-3-319-33024-2_12

started by the Swedish University of Agricultural Sciences (SLU) in the late 1980s. Landscape architects Roland Gustavsson from SLU and Carl Aage Rasmussen from Holstebro Municipality were two of the key people responsible for the initiative in Holstebro.

The first landscape laboratory of such a kind was established in 1986 in the SLU campus in Alnarp. There are currently landscape laboratories in Alnarp (Sweden), Snogeholm (Sweden), Holstebro (Denmark), and the new landscape laboratory being established by the Aarhus School of Architecture in Aarhus (Denmark).

Sletten is a visionary urban development plan integrating new housing with new urban forest, characterized by features often associated with urban gardening. The building plan comprises private and public housing, public institutions for children and eldercare, offices, and light industry. The housing areas consist of eight small clusters, each of which encloses a green area for leisure and play. The forest area is comprised of three primary experimental methods for afforesting and a number of subordinated experiments. They create numerous spatial and natural qualities that invite wide biodiversity and (just as importantly) wide variety for appropriating and using the urban forest. Compared to contemporary Danish city development projects, Sletten is very advanced in terms of new types of forest habitats as well as social diversity regulated by ownership of housing, inclusion of private and public institutions, integration of nature and city, and in urging citizens to take part in both nature development and preservation.

Integrated into the Sletten landscape laboratory is the Millennium Garden. This is a project inside the landscape laboratory project, a citizen-driven initiative directed at giving a landscape gift to future generations – from one millennium to the next – in the form of 55 oak trees planted in a square across both urban areas and the forest. The Millennium Garden is an example of bottom-up city and landscape planning.

In Sletten two different but interwoven stories can be read. The first story concerns the overall strategy of using Sletten as the basis for testing and developing new types of urban forest in close collaboration with the inhabitants. The other regards the many different kinds of citizen-driven forest appropriations (Fig. 12.1).

12.2 The Problem

During the last decades, development within both urbanism and forestry has led to the point that new forest is now created in the image of the city, insofar as forestry is increasingly dominated by urban values, laws, and demands (Konijnendijk 2008; 205). Since 1805, when a series of laws were implemented in Denmark to save the remaining forest from overcutting, forest areas in the country have risen from 2 % to approximately 14 % (Danish Nature Agency 2002). Following government plans from 1989, this afforestation will continue right into the twenty-first century, until one quarter of Denmark's land surface is forest area.

Until the mid-twentieth century, a forest was essentially a productive landscape, but there has been a pronounced change in its role in Danish and other Western

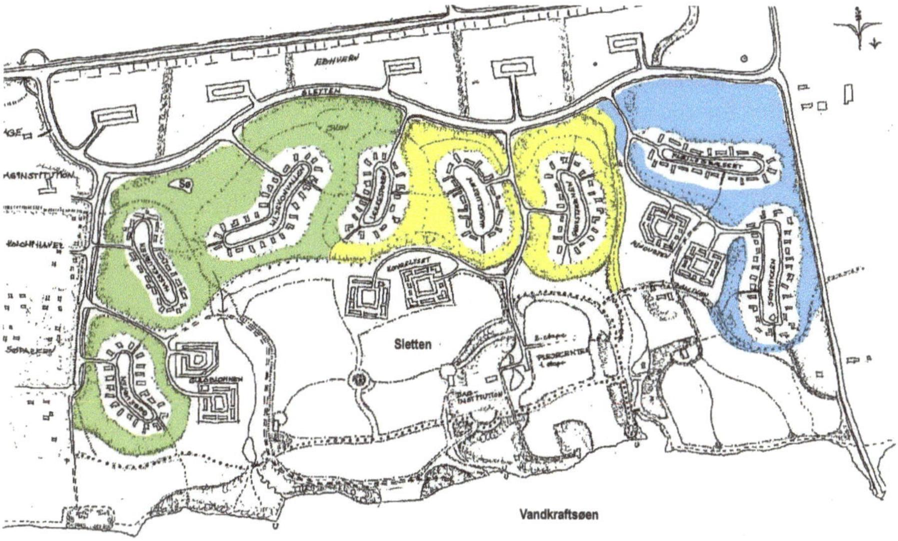

Fig. 12.1 The forest is comprised of three primary experimental afforestation models: habitat model (*green*), seed source model (*yellow*) and density gradient model (*blue*)

societies over recent decades. Today, afforestation in Denmark is closely linked to an increasingly urban society and is also justified in terms of recreational amenity. This development is related to the emergence of a new urban condition, in which the city can no longer be characterized as a demarcated domain surrounded by open countryside. It is increasingly becoming part of growing urban regions comprising both city and countryside (Sieverts 2008; 253). There is a rising awareness that urban forests can play an important role in creating an insight into the contemporary urban situation as a new, but as yet unrecognized, part of our everyday landscapes. Urban forests can also reconnect society and nature in places where we work and live (Konijnendijk 2008; 123). Along these lines, afforestation and urban gardening are also part of the discussion on nature and natural habitats in urban contexts. Sletten exemplifies that it could play an important role in qualifying urban development.

It is widely recognized that urbanization which has not yet reached its peak poses challenges for natural habitats in urban areas. In an attempt to rise to this challenge, cities such as Nantes, Lyon, and others are adopting strategies that include an increased focus on natural habitats in urban areas. Such strategies are also related to mitigating climatic changes, and their success criteria are widely dependent on civic engagement. In these perspectives, new types of urban afforestation, capable of merging landscape and dense urban areas, are becoming part of new urban landscapes. They merge natural habitats, urban development, and human engagement in taking care of and using natural features as a part of the same overall strategy, helping create and establish resilient urban environments.

Sletten is both an actual suggestion for novel neighborhood design and a full-scale experimental platform, in which scientists and practitioners with different or no professional backgrounds can meet and collaborate on testing and developing new concepts for the design, establishment, and management of urban forests (Gustavsson et al. 2005; 1).

Because Sletten is a living laboratory, there are many actors involved. The most important ones are Holstebro Municipality; the people living in Sletten (as they are encouraged to actively take part in the continued management and change of the urban forest); the participating researchers and research institutes comprising SLU, Copenhagen University, and the Aarhus School of Architecture; as well as both public and private actors in forestry and civic-culture sectors that are involved in urban afforestation and urban forest management.

The two main reasons for Sletten's establishment are closely linked to the two different stories that are unfolding in the area. The first is the question of diminishing resources available for the management and care of urban forests and green areas in general. In Sletten, this question was asked: can new types of civic-based management be part of the solution for the maintenance of new urban forests? This story would at the same time offer an answer to the second reason related to public and land use. In Denmark, the allotment garden movement has diminished over the last decades, but at the same time, there has been a growing demand for other types of urban gardening and the ability for people to settle themselves into their surroundings.

The people living in Sletten have been actively involved in the development from the very beginning. Curiously enough, this has led some of Sletten's inhabitants to establish grazing guilds and animal husbandry, contributing to maintaining the open grasslands despite the lack of any farming experience and facilities.

12.3 The Strategy

First and foremost, the project is spun out of a collective passion (shared by inhabitants, scientists, and the municipality) for creating a unique place built on the participation of the people living in the area from the very first day. One simple but effective tool fostering active participation and the acknowledgment of this has been to incorporate what initially was a so-called collective zone, a 3-m area between the various private gardens and the forest. Here inhabitants are encouraged to use the forest as they see fit. In several cases, this zone has expanded further into the forest, and a large number of temporary and self-organized forest gardens have emerged (Fig. 12.2).

Example 1
An inhabitant has been creating a bamboo nursery where the collective zone is used as a bamboo plant station and test field. The bamboo nursery has slowly become part of the specific area of forest directly related to where he lives, making not only the garden but also the forest an integrated part of his everyday landscape.

Fig. 12.2 Examples of appropriation

Example 2
Drawing upon the allotment garden movement, another family decided to create an enclosed garden in the forest and in turn establish both a flower garden and a vegetable plot. This particular area is an example of how many inhabitants of Sletten interact with the forest and, on the basis of this interaction, relate with the natural processes occurring in the forest over time.

Example 3
At one time, a group of inhabitants from one of the clusters decided upon a common community celebration and used the collective zone and the related part of the forest as the backdrop for their celebration. Such celebrations are common in Denmark and are usually referred to as "road parties." This in turn changed that particular area of forest into a new habitat for other types of plants that are growing in the area over time.

Sletten is a collective project and supported by a rare breed of visionary politicians, active citizens, and strong professionals, but there would be no such thing without a key person working in difficult times as well. Carl Aage Sørensen, former head of city landscape planning and infrastructure in Holstebro, is today a self-employed landscape consultant. In the very early stages of the project in 1997, Carl Aage Sørensen recognized the necessity to have Sletten work on a different basis from an ordinary park. Based on his knowledge of the establishment of landscape laboratories in Sweden, Roland Gustavson, the Swedish initiator, was invited to join the Danish experiment. Together with Christen Nørgaard, who impersonated the project contractor, the three became the core-connecting team of the project, ensuring political and public involvement in the project all the way through.

In the establishment of Sletten, the main actors have been Holstebro Municipality and SLU which, together with the inhabitants, are continuing to create and develop new management-oriented and experimental initiatives.

Furthermore, integrated into Sletten is the Millennium Garden, which was started by the Danish land artists Jette Hansen Møller and Erik Skoven (Copenhagen) in 1998, at the turn of the millennium. The Millennium Garden is a land art project, which was set up in relation to the establishment of Sletten and has become an integrated part of the landscape laboratory. It is an artist-based initiative directed at giving a landscape gift to future generations – from one millennium to the next – in the form of 55 oak trees planted in a square across both urban areas and the forest.

12.4 The Solution

Sletten has an overall forest structure with eight integrated clusters of various types of housing in close relation to the forest. The area covers 160 ha, which includes the overall structure of the forest as well as green wedges, village greens, and local gardens between individual houses and the forest.

Sletten was established in three phases and is based on three different afforestation models. The first phase comprises 36 smaller forest lots, each of 3,500 m^2. It is based on the *habitat model*, where every lot has a distinct composition of species, selected from around 60 in total. The second phase is made up of three types of basic woodland, oak, birch, and pine. It is based on the *seed-source model*, where an arrangement of seed-spreading bases with nine different tree species and nine different bush species will, over time, spread into the surrounding woodland to create new, self-organized forest stands of different species. The third phase is based on the *density gradient model*, where different distances between the used species create a large spatial variety. The trees are planted on grid patterns that vary in size from 1.25×1.25 to 2.5×5 m.

All three models incorporate the previously described "collective zone," where the inhabitants are encouraged to use the forest as they see fit. Despite the fact that the Department for Technique and Environment in the Municipality of Holstebro is responsible for the management of Sletten, it is to a large extent based on the still growing initiatives implemented by the inhabitants themselves. Additionally, workshops run by SLU are occasionally carried out in Sletten using one of the common houses as a departure point for projects in the field, in which the inhabitants are included. Martin Meisler Elmholdt-Svendsen has taken over from Carl Aage Sørensen as municipality-based contact person.

The "collective zone" and the self-organized gardens show Sletten is a space of potential, open to change over time, transforming into a mosaic of forest and garden habitats in continuous development and decline, with intentional and coincidental spaces of interaction in an increasing number of spatially open, half-open, and closed areas. In Sletten, temporality and transformation can be seen as an expression of the relationship between natural and cultural processes. While the forest itself is more permanent, the integrated self-organized gardens function as individual "experimental spaces" for the inhabitants and have a more temporary character. Despite the forest's planned structure, it is open to self-organization.

The situation originally imagined has been achieved. Sletten today is an urban forest maintained by the people living in the area – nature is a part of the life of people living there and so is the responsibility of keeping it. The municipality has acquired more forest and areas for leisure activity without great maintenance cost, and, in addition, they have fostered a civic culture in the neighborhood that shares pride in the area.

12.5 Lessons Learned

The experiments carried out in Sletten make the past, present, and future connections between individual human behavior, collective identity, small-scale complex systems, and ecological processes become visible. They show how urban forests like Sletten can (re)connect society and nature:

> In the depths of cultural memory forests remain the correlate of human transcendence. We call it the loss of nature, or the loss of wildlife habitat, or the loss of biodiversity, but underlying the ecological concern is perhaps a much deeper apprehension about the disappearance of boundaries, without which the human abode loses its grounding. (Harrison 1992; 246)

As the citizens learn from Sletten, our practices and language change and our daily rhythms shift. The nature of aesthetics and ethics shift toward what Leach describes as "open to the texture as much as to the text of everyday life" (Leach 2005; 141).

In line with the achievements in Samsø (Denmark), which through civic engagement and collective force has developed an energy self-sufficient and CO_2-neutral island (Hermansen and Nørretranders 2013), and the work of Elinor Oström (1990), which points to the superiority of collaborative over competitive behavior when exploiting shared resources, Sletten also points to the concept of "commons" as strong arenas for ecologically, culturally, economically, and socially sustainable development. What is achieved in situations like Sletten is what the creators behind the Energy Academy at Samsø call "commonity" – a contraction of "community" and "common." The contraction denotes that a "commonity" is a shared common resource maintained and kept in balance by obeying to collective, community-based rules governing the resource. Keeping the grasslands at Sletten in balance and thriving demands that every single actor with sheep, cows, goats, or other herbivorous animals does not send more animals than agreed into the fields. Imagine if a single individual stepped out of the commonity, maybe aiming at personal gain in the shape of more meat – the system's robust yet fragile ecological balance will slowly deteriorate, the field will stop being green, and the animals will all starve.

It takes little imagination to transfer the concept of commonity to other domains such as fresh air, clean water, good education, etc. It will take many Sletten and Samsø examples to change our everyday habits. However, Sletten and other similar initiatives help keep such otherwise abstract consequences tangibly alive at our doorstep and remind us of the robust yet fragile system we are all a part of and which no technology can mend without strong human action.

References

Danish Nature Agency (2002) Danmarks Nationale Skovprogram. Danish Nature Agency, Copenhagen

Gustavsson R, Nielsen A, Rasmussen CA (2005) Nye modeller for bynære blandingsskove. Videnblad 3:1–53

Harrison PR (1992) Forests: the shadow of civilisation. The University of Chicago Press, Chicago

Hermansen S, Nørretranders T (2013) Fælledskab = fælled + fællesskab. Samsø Energiakademi, Samsø

Konijnendijk CC (2008) The forest and the city. Springer, Berlin

Leach N (2005) Less aesthetics, more ethics. In: Ray N (ed) Architecture and its ethical dilemmas. Taylor & Francis, London, pp 135–142

Ostrom E (1990) Governing the commons: the evolution of institutions for collective action. Cambridge University Press, Cambridge

Sieverts T (2008) Improving the quality of fragmented urban landscapes – a global challenge! In: Seggern HV, Werner J, Grosse-Bächle L (eds) Creating knowledge – innovation strategies for designing urban landscapes. Jovis Verlag, Berlin, pp 252–265

Chapter 13
Milan: Community Gardens as a Space for New Societal Assemblages and Learning on Public Goods

Valeria Fedeli

Abstract This chapter presents the case of Milan as an example of Human Smart City referring to the urban agriculture experience. It discusses the process through which the Municipality of Milan has recently started investing in forms of urban agriculture as a way to enhance the active participation of citizens in the production of public goods and to promote unexpected solutions to unsolved urban problems. Urban agriculture in Milan, in this perspective, seems to provide an opportunity for reconciliation between the urban and nonurban realm in a densely built, urban environment and the possibility of introducing new approaches to public policies, urban planning included.

Keywords Urban agriculture • Community gardens • Urban planning and policies

13.1 The Context

More than 200 vegetable gardens have been registered and mapped in the city of Milan by a recent research project (Cognetti et al. 2014), covering a total surface of almost 1.7 million square meters. Mainly found in abandoned places, residual areas, and forgotten patches in the territorial figure of a city characterized by high building density and imperviousness, they account for less than 1/100 of the total urban surface. Fragmented and episodic spaces, these gardens speak of a sort of "backyard" city, the hidden face of a highly urbanized context, where the space for urban agriculture has been for decades residual or denied. Many of them are situated along mobility infrastructures, public open spaces and parks, or inside or along former industrial areas that have not been reclaimed yet. Only 20 % of them are based on formal agreements and have a legal status. Most of them, indeed, are in

V. Fedeli (✉)
Department of Architecture and Urban Studies, Politecnico di Milano, via Bonardi 3, 20133 Milan, Italy
e-mail: valeria.fedeli@polimi.it

G. Concilio, F. Rizzo (eds.), *Human Smart Cities*, Urban and Landscape Perspectives 19,
DOI 10.1007/978-3-319-33024-2_13

a condition of more or less explicit informality or illegality, although this situation seems to have slightly improved during the last two decades. In the nineties, only 10 % of existing vegetable gardens were in a condition of formal legality. This can be interpreted as the outcome of a recent process of convergence between the logics and rationales of insurgent social practices (which characterized the last decades) and a new interest and attention provided by the public administration, which has matured in the last years. The public administration only recently started investing again in forms of urban agriculture as a way to enhance the active participation of citizens in the production of public goods and to promote unexpected solutions to unsolved urban problems. Urban agriculture in Milan, in this perspective, seems to provide both an opportunity of reconciliation between the urban and nonurban realm in a densely built urban environment and a possibility of introducing new approaches to public policies, urban planning included.

The moment in which this happened was indeed particular. In 2010, a new municipal government was elected, headed by the Mayor Giuliano Pisapia, the party-independent leader of a center-left coalition supported by civic associations. After decades of center-right government and limited attention to civic society, he promised to promote a new approach to the government of the city, based on the strong involvement and empowerment of the citizens' movements and activists' networks that had supported the elections. Civil society, on the one hand, was suffering from the lack of ways and opportunities of involvement in policy-making and decision-making. On the other hand, it was characterized by a strong tradition – and interesting results – of self-organization in the production of local public goods. As observed by scholars in the field, the context of Milan is in fact characterized by a strong capacity for social innovation, but is backed by a weak and strategy-blind institutional setting (Dente 2005), often unable to follow and support this capacity. Bridging this gap was, in this sense, one of the most interesting promises of the newly elected mayor. This is even truer if considering that, during this period, the city (together with its larger urban region) had been suffering from the effects of the economic crisis, producing an unexpected polarized social context (Ranci 2012). This critical situation in terms of traditional policy action capacity was the result not only of the lack of resources, but of the lack of political vision and capacity of action by the public sector as well.

13.2 The Problem

Urban agriculture and community gardens have become the focus of attention and expectations of many scholars in the field of urban planning and policies. According to authors like Fraser (1990, 1992), for example, community gardens can play the role of *counter-spaces*, spaces for citizens to challenge institutions that are not able to provide appropriate answers to their needs and to experiment new forms of production of common goods. As such, they are seen as spaces for the expression and practice of a new public urban debate. According to authors like Staheli (Staheli et al. 2002), on the contrary, they can become spaces of gentrification and *clubization*, i.e., closed spaces for privileged actors, to access facilities and green areas, and protected spaces in which special citizens produce privatized common goods. In this sense, they can become spaces of conflict between different publics (*ibidem*, pg. 198). However, maps of urban agriculture can also represent new forms of cooperation and new forms of urban conflict (Heynen 2003). In both cases, even when insurgent social practices of urban agriculture are supported by public actors, community gardens, in an *actor network theory* (Ernston 2013), seem to provide interesting examples of how contemporary society is established and reproduced, unlike traditional twentieth-century practices. In fact, they talk about new possible social assemblages, new spaces in which urban commons are produced. At the same time, as localized practices, often supported by specific groups of local activists, community gardens offer the opportunity to discuss the so-called scalability of similar experiences – in other words, to answer one of the issues at stake in this book: to what extent are they able to activate opportunities that can produce not only localized common goods for the actors involved, but, more in general, new models of production of public goods for the larger urban community?

In this perspective, community gardens might be seen as a possible solution to the limits of traditional public policies stuck in traditional policy design and implementation approaches. Here the central challenge is if, and to what extent, community gardens can play the role of "learning arenas" (Bendt et al. 2013; 19; Wenger 1998, 2000), soft spaces where new forms of engagement (Thevenot 1994) occur through situated and thin interaction.

In this direction, insurgent practices of community gardens that became active in the last decades in Milan seem to be dealing with different kinds of objectives and quite different problems related to the crisis of public goods production, in order to reduce the distance between institutions and civic society. In fact, we can distinguish no fewer than three different models, in relation to the different available definitions of the problem (Cognetti et al. 2014).

The first family of practices mapped in the Milan case focuses on the relationship between people and places: a number of cases deal with the recovery of large or small empty areas, neglected by public action or abandoned by private owners, producing degradation and insecurity and generating further uncertainty about the possibility of reintroducing them into the urban transformation cycle. Others focus on a wider objective, i.e., producing a new sense of community through the rediscovery of neighborhood spaces, especially those situated in problematic contexts, such as public housing estates in marginal zones, abandoned by public action, and contributing as such to the general degradation of the context. This is particularly the case of an initiative in the northern sector of the city, promoted by the association supporting a historical school park, Parco Trotter, since 2009. The creation of a community garden is seen as a tool to cultivate relationships between the inhabitants of a multiethnic neighborhood. In the southern sector, next to another problematic neighborhood, Corvetto, the *Associazione Piano Terra* has opened a community garden, in the backyard of a school, with the aim of promoting the establishment of neighborhood ties around a space dedicated to children from the neighborhood (Cognetti et al. 2014).

A second family of practices focuses on a vast amount of marginal, residual, fragmented urban spaces, not necessarily bound to a specific local community or neighborhood, often "no man's land," forgotten by the public administration. Activists, in this case, promote demonstrative actions in order to show their potentialities in the production of the city, often through events and non-ordinary practices, as a form of proactive protest against the crisis in public action. A clear example is that provided by the *Ortodiffuso* project. Promoted in 2009 by the agronomist Mariella Bussolati, it tries to monitor practices of urban agriculture all over the city. At the same time, it promotes a bridge between activism and the action of public administration, in order to feed the public debate. In this sense, the *Rete Libere Rape metropolitane-Ortocircuito* network of projects connects very different initiatives. A typical example is that of the *Landgrab* collective, which uses "guerrilla gardening" in order to promote new forms of public participation. Landgrab was established in 2005 as a tool to oppose a large urban transformation project promoted by a partnership of public and private actors (Cognetti et al. 2014).

A third family of practices refers to urban agriculture as an occasion to develop new social cohesion, by sharing interests and practices. In this respect, agriculture is just a pretext to develop new social ties, promote the integration of different communities, empower people, etc. Initiatives of this kind use urban agriculture with the aim of experimenting new ways of producing public goods. It is the case of projects like the one launched by the *Cascina Bollate* social cooperative, which proposes agriculture as a tool for integrating the prisoners of the Milano-Bollate prison into society. Prisoners have been able to collaborate with gardeners of the *Attraverso il Giardino* Association since 2007 and help grow and sell vegetables. Urban agriculture is, in this perspective, a space of dialogue between the prisoners' world and the outside. It is also the case of the *Coltivando* project, proposed in the Politecnico di Milano University Campus in Bovisa and promoted by students and professors in 2012. The 600 m^2 area is open to university staff, to students, and to the local community as well, with the aim of trying to establish a new dialogue between the university and the city (Cognetti et al. 2014).

13.3 The Strategy

In all the cases explored, urban agriculture insurgent practices seem to look for new possible spaces for the spontaneous production of new "societal assemblages" (Latour 2005), "learning arenas" (Wenger 1998, 2000) and *commons* (Linn 1999; Eizember 2012b), and their potentialities in terms of public policies. To date, they have acted as counter-spaces, facing absent institutions unable to address the new urban question within a traditional definition of people's needs and policy tools (Fraser 1990, 1992). However, more recently they seem to have become opportunities for the crisis-stricken public sector to develop new approaches in order to produce public goods.

On the basis of the success and diffusion of these experiences, the municipal administration has shown an interest in exploring the potentialities of these insurgent practices as well.

In 2012 two laws issued by the Municipal Council Board opened a new "official season" for urban agriculture in Milan. The first law, adopted in September 2012 by four of the city's departments (Green Department, Decentralization Department, Department of Culture, and Department of Urban Planning), is part of a larger project called COLTIVAMI. The public administration offered citizens 25,000 m^2 to be assigned (after a public procedure) to associations or public and private actors, to be run as spaces for activating citizens so they would promote intergenerational and interethnic exchange. They were also to support the production, knowledge, and interest in environmental resources and to support the valorization of fragile land and spaces. Areas are assigned for 9 years free of charge, in order to generate "spaces for collective gardening able to favor socialization and social cohesion." With this project, the public administration, owner of more than 700 agricultural allotments in the city, aims at promoting individual micro-practices of sustainability able to feed new sustainable behaviors and promote area maintenance, in particular in fragile and underused areas. Community gardens are seen as spaces to experiment new public-private partnerships, possible forms of decentralization of public action. This strategy is supported by the new land-use plan that defines community gardens as public facilities. In order to become responsible for community gardens, actors have to produce a project, showing the way in which they will be able to involve citizens, together with a business plan and an activity program.

In a previous law, promoting the project "Giardini Condivisi" a few months earlier in the same year, the public administration promoted the involvement of citizens in processes of urban regeneration for unsolved, neglected, or problematic urban spaces. In this case, the space is assigned to associations on the basis of a project to be implemented for a minimum of 1 year to a maximum of 3 years. In this project, the administration acknowledges that specific urban spaces have public value and asks citizens to promote new redevelopment processes, able to achieve multiple objectives, from spatial renewal to social cohesion. In order to manage these spaces, actors have to establish nonprofit associations and present a master plan for the construction of the community garden, as a meeting space, able to favor socialization processes (Cognetti et al. 2014).

In both cases, the public administration defines a set of rules and objectives associated to urban gardens and proposes a new interpretation of urban gardens, from areas of individual practice for special types of people (elderly people, unemployed, disadvantaged people) to spaces for collective practices. At the same time, one should notice that the rules and procedures required, based on the preparation of a project and the constitution of citizens into an association, define quite a formalized mode of activation. This sounds a bit traditional and risks reducing the space of light innovation of the insurgent practices that are inspiring the acts of the public administration. Evidently, the public administration feels the need to produce a set of rules, which should guarantee the logic of public action. However, this can become a relevant obstacle for this kind of initiatives. In fact, the two laws issued by the public administration require a higher form of engagement from citizens that goes well beyond individuals and enters into the complex sphere of the intentional production of public goods. They follow ex post and try to both support and regulate the capacity of self-organization shown by civil society.

For the urban garden to become a space for shared practices, as suggested by Turner (1994; 120–130), a number of special conditions should occur and be present. Urban gardens and community gardens are the "eventual" meeting point of different "regimes d'engagement" (Thevenot 1994; 56–73) suspended between "regimes of familiarity," "regimes of regular action," and "public regimes of justification," since the space of a community garden is the space for autonomous and free action, as well as the space of intentionality, where the public sphere can eventually be present. In this respect, the expectations of the public administration could remain unfulfilled and the gardeners frustrated.

13.4 The Solution and Lessons Learned

The outcome of these two municipality laws is that a number of initiatives have recently started flourishing all over the city. Being quite new and still developing, it is actually quite difficult to discuss their results. In a former parking lot in via Montello, next to an army barrack, and in a public area in via Pepe, next to the Garibaldi railway station, in two central urban areas, two spaces have become successful public spaces, thanks to the "Giardini Condivisi" initiative. In particular, the second, officially opened in May 2013, through a public agreement between the association Isola Pepe Verde and the municipality of Milan, is playing an important role of socialization in the neighborhood (Cognetti et al. 2014). This example is actually a sort of intermediate one between the new opportunity provided by the public administration and the previous insurgent forms, because it is tightly linked to former activation processes in the local neighborhood, which opposed a large renovation project. A number of similar cases are now under experimentation throughout the city, but the outcomes cannot be taken for granted.

In fact, with Turner (1994; 120–130), when looking at community gardens as shared practice, we should ask ourselves *what* is shared – in other words, if this practice can be seen as a clear object or as a sum of individual and separate elements (Barnes 1994; 22), individual habits that coordinate, adjust to each other, and do not simply sum up together in order to produce a common objective in a traditional public action model. In this perspective, community gardens should be regarded as spaces for free, autonomous individual practices that can eventually cross elements of public relevance. A vegetable garden can become a community garden, because new trading zones could be established around it, where, as Galison said, "*trade* focuses on coordinated, *local* actions, enabled by the *thinness* of interpretation rather than the thickness of consensus" (Galison 2010; 36). In this perspective, the practices supported by the public administration can be read as a contradictory and challenging trial to support ex post the capacity of organization and innovation of social practices, or as an open space for collective learning for both public administration and individual actors, thin practices producing new small learning arenas around problems of public interest. In this sense, they can help notice and deal with the difficulties of renovating traditional public action forms.

On the one hand, they offer the opportunity to consolidate into practice the fragile principles on which urban sustainability can be built (Callon et al. 2001) allowing citizens to become central actors in the construction of a new environmental consciousness. On the other, community gardens do not exist out of the community of practices that feed them, and their potentialities cannot exist out of the relationship with the community that produces them – the *commoners* (De Angelis 2003, 2007). Community gardens, like commons, should be regarded and promoted as the outcome of an action that produces goods that do not enter in the sphere of *commodification* (Eizemberg 2012a). There are no commons without "commoning" (De Angelis 2003, 2007). Public administration should concentrate its efforts to foster "commoners" and "commoning," but the challenge is still quite consistent and far from being achieved.

References

Barnes B (1994) Practice as collective action. In: Schatzki TR, Knorr Cetina K, Von Savigny E (eds) The practice turn in contemporary theory. Routledge, London

Bendt P, Barthel S, Colding J (2013) Civic greening and environmental learning in public-access community gardens in Berlin. Landsc Urban Plann 109:18–30

Callon M, Lascoumes P, Barthe Y (2001) Agir dans un monde incertain. Essai sur la democratie technique. Seuil, Paris

Cognetti F, Conti S, Fedeli V (2014) La terra della città. Giardini coltivati e giardini condivisi a Milano. In: Ferraresi G (ed) Il Progetto di Territorio. Maggioli, Milano, pp 113–168

De Angelis M (2003) Reflections on alternatives, commons and communities or building a new world from the bottom up. Commoner 1–14. http://www.commoner.org.uk/deangelis06.pdf

De Angelis M (2007) The beginning of history: value struggles and global capital. Pluto Press, London

Dente B (2005) Governare l'innovazione a Milano. Impresa e Stato, 71:1–4. http://www.mi.camcom.it/upload/file/1235/617611/FILENAME/11Dente.pdf

Eizenberg E (2012a) The changing meaning of community space: two models of NGO management of community gardens in New York City. Int J Urban Reg Res 36:106–120

Eizenberg E (2012b) Actually existing commons: three moments of space of community gardens in New York City. Antipode 44(3):764–782

Ernstson H (2013) The social production of ecosystem services: a framework for studying environmental justice and ecological complexity in urbanized landscapes. Landsc Urban Plan 109:7–17

Fraser N (1990) Rethinking the public sphere: a contribution to the critique of actually existing democracy. Social Text 25(26):56–80

Fraser N (1992) Rethinking the public sphere: a contribution to the critique of actually existing democracy. In: Calhoun C (ed) Habermas and the public sphere. MIT Press, Cambridge

Galison P (2010) Trading with the enemy. In: Gorman ME (ed) Trading zones and interactional expertise: creating new kinds of collaboration. MIT Press, Cambridge

Heynen NC (2003) The scalar production of injustice within the urban forest. Antipode 35: 980–998

Latour B (2005) Reassembling the social: an introduction to actor-network-theory. Oxford University Press, Oxford

Linn K (1999) Reclaiming the sacred commons. New Village 1:42–49

Ranci C (ed) (2012) Città nella rete globale. Bruno Mondadori, Milano

Staeheli LA, Mitchell D, Gibson K (2002) Conflicting rights to the city in New York's community gardens. GeoJournal 58:197–205

Thevenot V (1994) Pragmatic regimes governing the engagement with the world. In: Schatzki TR, Knorr Cetina K, Von Savigny E (eds) The practice turn in contemporary theory. Routledge, London

Turner S (1994) Throwing out the tacit rulebook: learning and practices. In: Schatzki TR, Knorr Cetina K, Von Savigny E (eds) The practice turn in contemporary theory. Routledge, London

Wenger E (1998) Communities of practice, learning, meaning and identity. Cambridge University Press, Cambridge

Wenger E (2000) Communities of practice and social learning systems. Organization 7(2):225–246

Chapter 14
Aveiro: Civic Movements to Promote Smarter Decisions for the Future of the City

José Carlos Mota and Gonçalo Santinha

Abstract Aveiro is the second largest city in the Centro Region of Portugal. In 2009, a major urban regeneration project named 'The Sustainability Park' (*Parque da Sustentabilidade*) was launched by the city council. Supported by the National Strategic Reference Framework (NSRF 2007–2013), the project involved 15 local and national partners, including the University of Aveiro and the Association of Small Businesses (commerce), and a budget of nearly 14 million euro, subdivided into 17 subprojects. This chapter presents the story of the project. As soon as the project became public, primarily through the media, citizens protested about the fact that they had not participated in the proposal's design, nor had they been informed, to say the very least. This situation prompted a strong civic movement to call the City Council's attention to the potential harmful impact of these projects on the local environment and on the daily lives of residents. This contribution describes how movements tried to engage in a series of talks with the politicians involved and bring tacit and codified knowledge to the process.

Keywords Urban regeneration • Public space • Design for sustainability

14.1 The Context

With approximately 55,000 inhabitants, Aveiro is the second largest city in the Centro Region of Portugal. As the capital of the Aveiro inter-municipal community subregion, a large number of administrative, cultural and health services are concentrated in this city. It is the home of the University of Aveiro, attracting thousands of students from across the country and standing out as a catalyst for local and regional development. Often described as the Venice of Portugal for its canals and boats, Aveiro is usually seen by tourists and local residents as a good place for walking and cycling because there are no hills.

In 2009, a major urban regeneration project named 'The Sustainability Park' (*Parque da Sustentabilidade*) was launched by the City Council. Supported by

J.C. Mota, Ph.D. (✉) • G. Santinha, Ph.D.
Department of Social, Political and Territorial Sciences, University of Aveiro, Aveiro, Portugal
e-mail: jcmota@ua.pt; g.santinha@ua.pt

G. Concilio, F. Rizzo (eds.), *Human Smart Cities*, Urban and Landscape Perspectives 19,
DOI 10.1007/978-3-319-33024-2_14

Fig. 14.1 Parque da sustentabilidade

the National Strategic Reference Framework (NSRF 2007–2013), the project involved 15 local and national partners, including the University of Aveiro and the Association of Small Businesses (commerce), and a budget of nearly 14 million euro, subdivided into 17 subprojects (Fig. 14.1).

The initiative came forth with the aim of designing and implementing a greenway that crossed a significant part of the city centre (199,106 m^2). By doing so, a set of facilities would become more integrated, and an innovative environment for residents and tourists would be promoted under the umbrella of the sustainable development concept. In the project report's own words, it was intended to 'affirm the city as a place of innovation, competitiveness' through a spatial intervention that residents and visitors would view with renewed interest.

Overall, the project included the redevelopment of an entire green area surrounding several old neighbourhoods and the creation of a public space that allowed citizens to be in close contact with 'nature' and good environmental practices. From a conceptual viewpoint, the idea was to promote sustainability-oriented interventions based on the ample notion of 'Design for Sustainability' (Birkeland 2002). Two elements were thus considered of paramount importance in the project design. The first one was the motto: sustainability. This implied that the project envisaged more than the physical rehabilitation of a set of green areas. In fact, it also looked at how leisure facilities were built in an integrated manner in green areas, how their design and management could promote energy efficiency and finally how the greenway could be tied up with the surrounding environment. The second element was the partnership created for the proposal. As the University of Aveiro has a long tradition in working with environmental and natural resource issues, with particular focus on sustainability issues, their knowledge was considered particularly useful in contributing to a more effective and reasoned proposal for action, as required by the funding support mechanism.

The initiative also envisioned to be a good practice role model to be replicated not only in other areas of the city but by other City Councils as well. According to the project promoters, the motto and the method used (collaborative) were an added value 'to promote the quality of the city environment and the welfare of its citizens and hence should be replicated in the future'. Bearing this in mind, the promoters proposed producing a good practice learning document called 'Aveiro 2020 – A Sustainable Strategy'.

14.2 The Problem

Such a complex initiative crossing an important and extensive area of the city centre soon engendered a set of problems. To begin with, the greenway concept comprised questionable spatial planning options. The construction of a pedestrian bridge over the city's central canal and the renovation of an old traditional garden, without a spatial planning framework and a clear understanding of the key issues to be addressed, are two examples of such debatable options. Secondly, no mechanisms for public scrutiny over the project design proposals were planned. Public participation was thus seen as a mere bureaucratic requirement. Finally, there were several operating difficulties in the notion of 'Design for Sustainability' – more than a new approach concerned with environmental, social and economic outcomes, such as energy efficiency, changing mobility patterns and improving social interaction, the project implementation actually emphasised a traditional urban approach with a 'green label'.

Despite being a project with a relevant theme – promoting sustainability in the context of urban regeneration – the Sustainability Park soon proved to have spatial planning proposals which were not coherent with the project's overarching objectives. At first sight, it seemed that the rhetoric used in the project's proposal was just an excuse to obtain the NSRF financial support, which demanded the combination of infrastructural interventions with the development of social capital, environmental protection and the promotion of networks and functional interlinkages. Accordingly, the complexity of the proposed exercise was largely neglected, as the project was basically understood to be an old-fashioned public space renovation and, as such, neither overcoming traditional policy 'silos' nor understanding the reasoning behind decision making in a cross-sectoral way.

The theoretical justification of the project proposals also lacked evidence. For example, the pedestrian bridge with high visual impact was justified by 'a long-felt need' that such a structure would allow a very large number of citizens who usually crossed the central canal to save considerable travel time. However, not only no research on pedestrian flows able to support the proposal was presented, but some empirical evidence (produced in the meantime by a group of spatial planners) actually contradicted the assumption. Another example is the proposed destruction of an old traditional neighbourhood garden in order to build, across the neighbourhood, a road to improve traffic flow. The argument presented by the City Council was the need to improve the design of the public space, enlarge the area of the garden and decrease the number of parking lots. Again, no studies to support the decision or to evaluate its consequences were presented.

As soon as the project became public, primarily through the media, citizens protested about the fact that they had not participated in the proposal's design, nor had they been informed, to say the very least. This situation prompted a strong civic movement to call the City Council's attention to the potential harmful impact of these projects on the local environment and the daily lives of residents. Despite their efforts, which included the submission of alternative solutions (especially regarding the two examples mentioned above), the civic movement's claims were successively ignored. From the point of view of the project coordination, there was no interest in justifying the solutions adopted and discussing possible alternatives, which would postpone the project's implementation.

One of the arguments presented by the civic movement was that all these conflicts could have been avoided if a different methodological approach had been adopted in the project's proposal design. One of the reasons behind the behaviour of the civic movement participants was the idea of preventing future policy decisions of this importance being made in such a centralised way. For example, although legislation does not include a public consultation procedure for this type of project, even if the own funding mechanism values and encourages this approach, the project could have included procedures for public participation and discussion with the community and, in this way, made people aware of the proposed solutions.

Another argument was the absence of a well-founded conceptual frame of reference common to all the involved actors, which could be used as a guide for citizens and other stakeholders to understand the project's scope. In fact, the development of such conceptual reference could be seen as a way to address the complexity of the new approach demanded by the financial support mechanisms and, at the same time, to seek synergies amongst the different stakeholders involved in the process. Being a project based on the motto of sustainability, with a strong emphasis on environmental issues, the participation of partners such as the University of Aveiro was seen as an added value in the process. In the civic movement's opinion, however, several examples of dubious interpretation of the concept could be found in the project's proposal, both in its social element (the lack of respect for history, identity and memory of old neighbourhoods and central canal) and environmental element (destruction of the existing tree structure in an ancient garden).

14.3 The Strategy

The ideas underlying the project's proposal ('Design for Sustainability') and the methodological principles of action (inter-institutional partnership) provided the ingredients to overcome the challenges linked to such a new approach demanded by the financial support mechanism. However, as soon as the project became public and was to be implemented, a huge debate began to take place in the local community. The emergence of the civic movement referred to above, *Amigosd'Avenida* (friends of the city avenue), was the community's driving force to counteract the problems

Fig. 14.2 Public meetings

detected in the project's proposal. It was joined shortly after by other civic movements in the city. One was the *Plataforma Cidades* (cities' platform) led by a well-known local architect, Pompílio Souto, who promoted a series of talks and discussions about the aims of the Sustainability Park. Another one was the *Comissão de Moradores do Alboi* (Alboi Residents Committee), which focused its attention on the Alboi neighbourhood proposal. Finally, the *Movimento Cívico Por Aveiro – Contra a Ponte Pedonal* (Against the Footbridge) dedicated its attention to the foot bridge over the central canal, which citizens found to be a negative environmental and social intervention. Established as informal and nonpartisan, without any level of institutionalisation, all these movements sought to give voice to citizens and create the conditions to improve not this process alone, but future decision-making processes as well (Fig. 14.2).

In unison, these movements tried to engage in a series of talks with the political power and bring tacit and codified knowledge to the process. Facing the City Council's several years of sustained opposition to explaining the project proposals and promoting a real public participation process (with the argument that there was no time or that the project proposal had already been shown in local newsletters and newspapers), the level of discussion increased, and each movement developed its own agenda.

The first civic movement, *Amigosd'Avenida*, with regular activity in the last 5 years, organised two lines of action from the beginning. On the one hand, the movement developed a set of digital platforms, including a mailing list with over

300 members, a collective blog (with over 120,000 visits, 200,000 page visits) and the Facebook Group 'Aveiro 2020' (with more than 2000 members after 3 years of activity), which mobilised many citizens to the cause. On the other hand, the movement organised local community discussions to complement the digital discussions and the arguments presented there. In one of the public events, and because the City Council refused to attend, the citizens documented the conclusions reached and took them to the Municipal Assembly to be read to all present. In addition, the movement promoted a petition that involved more than 400 signatories to apply for an extraordinary Municipal Assembly to present and discuss the project proposals. Despite the twenty interventions over the project, the citizens were not allowed to counteract or to have any kind of dialogue with the project's technical leaders. As a result, none of the concerns of the citizens were taken into account, and the process moved forward without any changes.

Unlike *Amigosd'Avenida*, the *Plataforma Cidades* was not created as a result of the Sustainability Park project. With more than 10 years of existence, this movement organises monthly gatherings of researchers, businessmen, school teachers and members of civil society. However, as soon the protests began, this movement organised a public appeal to the city council and project partners, pointing out the strengths of the project, not questioning the quality and possible impacts of the proposals presented, but raising concerns about the financial viability of some proposals. The text was signed by two former ministers' university professors. Oddly, the appeal had no effect at all.

As for the *Comissão de Moradores do Alboi*, this movement was exclusively composed of neighbourhood residents, especially the elderly. Focusing on the defence of the old garden and neighbourhood characteristics, this movement spurred a number of activities to protest and call for a change in the proposal. Examples include the development of a blog and a Facebook page (with 2500 friends) and several local protests in the public space. The impact of these activities in the community had such a visibility that Joaquim Pavão, a known film director and musician, decided to produce a short film called *Alboi um canto do mundo* (Alboi in the Corner of the World), which turned out to be a success in social media, with appearances on the radio and national TV and with more than 5000 visits in the Internet. In addition, an alternative proposal for the garden and the neighbourhood was presented to the city council, but despite the efforts, the decision to build a road in the middle of the garden remained.

Movimento Cívico Por Aveiro – Contra a Ponte Pedonal was the civic movement that mobilised the highest number of citizens (40) and more active protests. In just a couple of months, this movement collected more than 3500 signatures against the construction of the bridge and organised four public meetings with more than 500 people. Again, the use of virtual networks (Facebook page and mailing list) was a key issue to ensure the information spread and to increase the number of members. The movement also produced a technical and juridical document, with the support of experts in spatial planning, appealing to the national government, regional authorities and courts. The protests had a huge impact in the local mass media and national TV. Once more, these activities had no impact on the City Council's decision to construct the bridge.

14.4 The Solution

The arguments described above and the citizens' social refutation of the project solutions and process exposed how the concept of 'Design for Sustainability' was not well understood by the project consortium. The relevance of the theme was clearly not enough to put into practice the challenges that a project of this nature and importance to the local community entail.

The lack of formal public participation mechanisms led the city civic movements to appeal to the national government, the financial support (regional administration) and the courts in order to solve conceptual and legal conflicts. This, in turn, increased the citizens' hostility towards the project and the alienation of political actors who had previously supported the project. In fact, public and semipublic (social networks) demonstrations of political actors were seen soon after, calling for an amendment of the proposals. In moments of political turmoil, the coalition members managing the City Council came into conflict and internal discussions conducted to a winning proposal from the opposition party to eliminate one of the most contested proposals – the design of a road crossing the garden of Alboi.

One aspect that deserves to be highlighted is that the increasing social contestation against the foot bridge, empowered by the legal and technical arguments presented in the courts (which claimed that the project was illegal because it was going against a previously approved formal urban plan), created a negative political context around the mayor and his political supporters.

This political instability had a strong impact in two different ways. Firstly, the project partners' relationship started to show signs of unsteadiness. For instance, some of the better known partners, such as the University of Aveiro and the *Associação para o Estudo e Defesa do Património Natural e Cultural da Região de Aveiro* (Association of Heritage), publicly expressed their disagreement with the whole process. Actually, this position clearly counteracts the idea of consensus and partnership that had prevailed from the beginning of the project and that was used by the City Council to support their argument for not looking at alternative solutions. Secondly, the company in charge of building the bridge over the central canal abandoned the process due to the process delays and legal doubts. After further unsuccessful attempts to contract another company, and in the face of financial problems due to the special nature of such intervention, the City Council decided to suspend the work on the grounds of 'seeking greater consensus'.

14.5 Lessons Learnt

The case presented in this section sought to show how the design and implementation of a project with important urban, social and environmental changes in a city gave rise to a strong reaction from the local community, challenging the mind-sets of politicians, stakeholders and citizens in general. The project, 'The Sustainability Park', had all the ingredients to become a good practice case to be replicated, not only in the future in other parts of the city of Aveiro but also in other cities, with the motto (sustainability), the partnership (collaborative environment with 15 institutions) and the challenges posed by the NSRF 2007–2013 (integrating sectoral and horizontal policies, emphasising functional relationships). However, a set of drawbacks clearly prevented the implementation of the project as foreseen by its promoters. They included the incapacity to manage the transition from a traditional approach (emphasis on the physical structure) to the new emerging needs (importance of intangible resources and a cross-sectoral policy perspective), the lack of a solid and common definition of the project aims based on the concept of 'Design for Sustainability' and accepted by all the consortium members and the scarce involvement of citizens from a very early stage of the process.

Many conclusions can be drawn from this case analysis. For the purposes of this chapter, three guidelines for public policy-making will be pointed out in a summarised way.

The first one is the importance of reinforcing cooperation arrangements between all the partners involved in the consortium, in order to promote a clear understanding of the project aims. To achieve this, it is essential that the project is codesigned by sharing objectives, understanding the nature of the challenges involved and the skills and experience necessary to embrace them. The development of a theoretical framework is essential to ensure a common understanding of the concepts and a solid support of the actions designed. In fact, this is a crucial step for what Morgan (2004) claimed as the transition from the power to decide and design policies to the power to transform or deliver those polices, which can only be achieved with the active cooperation of the local community.

The second one is the significance of using projects of this nature and dimension with a pedagogical purpose as well. This means that it is crucial to, firstly, engage and make citizens aware of the project design from the early stage (still in the diagnosis phase) and, secondly, to define the proposal in a joint manner (combining the efforts and solutions presented by the local community), not only because projects like this address collective challenges but also because they most likely involve behavioural changes (such as mobility patterns). Thirdly, the whole process is to be used to motivate citizens to participate in other similar processes and to discuss with them how to tackle the next challenges and the future investment priorities for local and regional development. In Portugal, the promotion of active dialogues between local governments and the local community is conventionally quite limited. Although local governments tend not to induce civic participation, the will of citizens to participate in such processes is also traditionally low. In the

'Sustainability Park' case, the fact that a lot was at stake, mainly for the residents, may partly explain the strong commitment of the local community. But this case also suggests that citizens do get involved in such processes when they have more access to information (expertise), when they have more knowledge about the problems at stake and possible solutions and when they notice that the community shares a common set of ideas and their points of view are taken into account.

The third guideline is the acknowledgement of social media as a tool to change policies and transform opinions into practical initiatives. The importance of technological innovative solutions to promote territorial governance is well documented (see *inter alia* Santinha and Castro 2010). Overall, the use of information and communication technologies supporting public participation can be part of a government's policy of openness and transparency. In the particular case of the project discussed here, its role can be expanded to a different level: the capacity to promote citizens' joint efforts to direct their community and encourage civic engagement (for further reading, see Mota and Santinha 2012; Mota 2014).

References

Birkeland J (2002) Design for sustainability: a sourcebook of integrated, eco-logical solutions. Earthscan Publications, London

Morgan K (2004) Sustainable regions: governance, innovation and scale. Eur Plan Stud 12(6):871–889

Mota J (2014) Planeamento do Território: Metodologias, Actores e Participação. PhD thesis, University of Aveiro, Portugal

Mota J, Santinha G (2012) Social media and civic engagement: discussing the case of Aveiro, Portugal. E-Pract 16:29–41

Santinha G, Castro E (2010) Creating more intelligent cities: the role of ICT in promoting territorial governance. J Urban Technol 17(2):77–98

Chapter 15
Palermo: Living Labs for Urban Regeneration

Ferdinando Trapani

Abstract The case of Palermo is described in this chapter as an example of Human Smart City (HSC) approach transferred to an extremely weak urban context in Southern Italy. The case focuses on the processes boosted through the Territorial Living Lab established in the city to tackle the topic of citizen participation in the solution of the main city challenges. It was based on a model of incremental and adaptive process of interaction, within a co-creative ecosystem, with the aim of improving housing quality. The model was shared by the players involved in the social innovation process. It would not have been possible without the Living Lab approach and the innovative momentum of design thinking translated into public practice.

Keywords Living Lab • Urban regeneration • Strategic planning • Participatory design

15.1 Foreword

An *electronic town meeting* (ETM) was held on February 12, 2012 as part of the framework of an experimental project on social innovation, which started in 2009 with the Territorial Living Lab Sicily (TLLS), the first Living Lab in Sicily. Other ETMs were organised in Palermo from 2013 to the end of 2014. The process underpinning the ETMs (Lukensmeyer et al. 2005; Aicardi and Garramone 2011)

The author was the scientific coordinator of the University of Palermo, partner under the European project Parterre which finished at the end of 2012 and was financed by the ICT Policy Support Programme, Competitiveness and Innovation Framework Programme 2007–2013 (http://www.parterre-project.eu/index.php). He took part in the organisation of five electronic town meetings promoted by the Department of Participation and Decentralisation of the Municipal Administration of Palermo from 2013 to the end of 2014.

F. Trapani (✉)
Department of Architecture, University of Palermo, Palermo, Italy
e-mail: trapanif@gmail.com

G. Concilio, F. Rizzo (eds.), *Human Smart Cities*, Urban and Landscape Perspectives 19,
DOI 10.1007/978-3-319-33024-2_15

in Palermo is an example of Human Smart City (HSC) approach transferred to an extremely weak urban context in Southern Italy.

The case study (Marsh et al. 2013) has a number of critical issues, mainly for the reasons listed below: (a) the TLLS experimented the HSC model in a time of continuous changes in the political class; (b) the city is developing its own urban planning and infrastructure planning tools; (c) the very first experience was carried out in the neighbourhood of Brancaccio, which is commonly known for the presence of the Mafia; and (d) the technical and administrative know-how of the municipality was poor, partly because of the lack of previous experience in institutional participation. All ETMs are aimed at achieving the highest possible levels of participation, based on known qualitative levels (Arnstein 1969; Bobbio 2002, 2008).

The university first (and the municipality of Palermo later) relied on the TLLS to hold the ETMs. The Parterre project focused on testing innovative models of strategic environmental planning and assessment in various European countries through the Living Labs. Unlike the prevailing notions of *smart city* (Hollands 2008; Greenfield 2013), the Living Lab is a travelling approach that can be understood as the use of technologies available to most users (widespread and most of all *social-driven*[1] technologies), because it has recently been the subject of investigation in some international cooperation projects (Marsh 2011; Concilio et al. 2012). The experimental participated event held in Brancaccio would not have been possible without the *Living Lab* approach and the innovative momentum of *design thinking* (Brown 2009) translated into public practice.

In Palermo, the TLLS tackled the topics of participation on the basis of a model of incremental and adaptive process interaction, within a co-creative ecosystem, with the aim of improving housing quality, which was shared by the players involved in the social innovation process.

The TLLS promoted a number of projects and social entertainment events on the Digital Agenda, while the city of Palermo (the seat of the regional government and a metropolitan city with more than 600,000 inhabitants) was the setting for a specific experience of Human Smart City (HSC), which focused on various urban topics (town plan, mobility, waste, coast, decentralisation) with the ETMs. This work is a scientific report on the TLLS's experience regarding the first ETM held in Palermo.

---

[1]For in-depth information about HSC, read the outcomes of the European project Peripheria (ICT PSP CIP programme) and in particular the document "Cook Book": http://www.peripheria.eu/library/human-smart-cities-cookbook. For information about the *human-centred design* approach, refer to IDEO "toolkit" available on the website: https://hcd-connect-production.s3.amazonaws.com/toolkit/en/download/ideo_hcd_toolkit_final_cc_superlr.pdf; last updated on September 29, 2014.

15.2 Background

The territory we refer to is municipal district no. 2, with more than 70,000 inhabitants, located south of the city centre of Palermo, in an area enclosed by the sea and the mountains that surround the city. It is divided into three heterogeneous strips of land that run parallel to the shoreline (the coast, the consolidated city and the countryside). Urban development has been following a different route since 1600, which caused the current district no. 2 to be excluded from any investments in services or buildings. Another reason behind its slow development over the centuries is the presence of the Oreto River. For many, many years slaughterhouses and tanneries were active on the banks of the river, which certainly did not contribute to the salubriousness of the surroundings.

The construction of economic and social housing over the last 40 years negatively contributed to the image of the area in terms of social isolation as well.

The presence of the underground railway link, the road connecting the sea to the mountains and, last but not least, the ring road contributed to making this territory even more fragmented and highly heterogeneous. Nonetheless, it is a territory rich in historical and architectural heritage, with the Castle of Maredolce, the villas and the *bagli* (typical buildings used for rural productive activities especially in Southern Italy) and extraordinary landscape. The coastline is potentially suitable for swimming, but at present it is in a state of total neglect.

In the past, Brancaccio played a major urban role, which has yet to be unveiled through archaeological and historical studies. The Maredolce palace stands on the former control point of the water supply used for the irrigation of highly fertile lands. The exceptional production of citrus fruits (in terms of quantity and quality) ensured high proceeds until the 1960s, which is why the area of Brancaccio and the current municipal district no. 2 are still free from buildings and dominate the rural landscape with the "gardens"[2] of Ciaculli, Croceverde, and the remaining cultivated lands in the eastern part of the city. When the palace was first built, the royal family and their court had been living in a special place.

15.3 Critical Issues

The critical issues experienced by Brancaccio do not only concern the long process of "regaining" the Maredolce palace for the public, which was triggered by the action of a single restoration expert (Braida 1965) who fought against the local Mafia alone.

The social, economic and urban planning problems of the neighbourhood of Brancaccio, which have been extensively investigated (De Spuches 2007;

[2]In Sicily, by *garden* it is usually meant a citrus orchard.

Next 2008), mostly emerged in the aftermath of the war in some areas and for specific reasons. Today, the area is in such a high state of decay that the palace, and what one day will be the park of Maredolce, can hardly be seen.

At present, the situation has changed due to the construction of the Forum shopping mall in an area that is connected to the perimeter of the industrial area. Unlike the fenced lots used for craft activities, it is open to the surrounding areas, and indeed it is the main attraction of Brancaccio for the neighbouring municipal centres. The residents first and the cultural and anti-Mafia associations later took a stand and laid the foundations for changes that are increasingly rooted even in areas and districts traditionally ruled by the Mafia. One example is the "MandarInArte"[3] project, at the heart of the suburb of Ciaculli, which is experimenting the relational model as the key to a new identity rooting, using premises and lands seized from the Mafia.[4]

The cultural associations of Brancaccio[5] are mostly responsible for the citizens' greatly renewed interest in the urban context, as well as the school association, which for some time now has been trying to protect the district's cultural resources. It is indeed these local strengths that have defended the castle and have been strongly committed, for many years now, to defending and enhancing the architectural, archaeological and environmental complex of the palace and former lake of Maredolce, which is famous in Italy and abroad.

15.4 The Strategy

Two years after the completion of the Parterre project, a number of cultural and social stakeholders still believe to this day that taking care of Maredolce means improving the life quality of the whole neighbourhood (Prescia and Trapani 2012; Prescia 2012).

A few processes of participated urban transformation have been implemented in Palermo. Among a number of experiences, two (the cultural sites of Zisa and the new Parco Uditore) are worth mentioning and draw public attention. In these cases technology contributed to the realisation of the two citizenship hot spots, although not to the same extent it did for the ETMs in Palermo, in which it gave a crucial contribution.

[3]For more information, see http://acunamatataonlus.weebly.com/progetto-mandarinarte.html, retrieved on September 20, 2014.

[4]For more information about the project, see http://www.mandarinarte.org/il-progetto

[5]Reference is made in particular to three associations involved in the activities carried out as part of the Medlab and Parterre projects: the cultural association "Castello di Maredolce" (set up in 1999); the Liceo Basile, based in Via S. Ciro in Palermo; and the Movimento di Promozione Umana (Human Promotion Movement) which is one of the Sicilian civic movements represented by the "Idea e Azione" Movement of Palermo. This group of local stakeholders asked the university to take care of the palace, the park and the Brancaccio neighbourhood.

15.5 The Solutions Adopted

15.5.1 Preparation Phase

By the end of 2010, a number of local associations had already started to consider some actions to be implemented in the Maredolce area and the whole district, i.e. accessibility, public and social housing, cultural heritage, security, tourism, social services, agriculture and urban vegetable gardens, waste and sustainable water use.

The method adopted for the preparation phase included visits to the neighbourhoods, interviews and sessions of "planning for real" along with the use of blogs and social networks.

Moreover, the Local Agenda 21 Organisational Unit of the municipality of Palermo held meetings for further exchange and dialogue. The activities mentioned above allowed to identify the strengths and weaknesses of the area, any perceived problems and the latent needs of the territory, such as housing quality, the integration of public institutions, counteracting social decay, urban congestion and the abandoning of residual rural areas (Trapani 2011a, b).

15.5.2 Outcomes of Participatory Experiences

A very short summary of the assessment report will follow. Concerning the outcome of the ETM, 80 % of the participants stated that they would take part in a similar initiative again, and 77 % said they fully understood the goals of the ETM. Other positive considerations were expressed about the use of electronic technologies and the discussions held sitting around the same table. In this respect, most participants judged the ETM to be a good instrument to find solutions to local problems, to define the guidelines to be implemented in the local development strategic plans, as a "databank of ideas" for future interventions and, in general, as a method to find potentially binding solutions for future choices.

The outcome of the ETM is, firstly, a real social document obtained in a way uncommon for the city (Giambalvo et al. 2013) and, secondly, a document of useful programmatic contents to define the general guidelines of the new general town plan and for the strategic environmental assessment.

15.5.3 The New Cycle

The European project Parterre ended with two public events. The first took place in the university premises of the rectorate of Palermo and included the official delivery of the Instant Report, prepared by the ETM assembly in Brancaccio in February 2012, to the new president of the municipal district. The second event concluding the Parterre project was an exhibition of graduation theses and lab projects carried out by many university specialties (urban planning, restoration and architectural design) based on the interpretation of the Instant Report's contents.

At the end of 2012, the municipal council for participation and decentralisation published a call for proposals that allowed some TLLS members – as an external body to the municipality – to organise and hold five ETMs in total, the last of which focused on the new general town plan.

15.6 Lessons Learned

The clear outcome of the first ETM obtained and executed by the TLLS is the recovery and updating of the administrative decentralisation policies. At last, today, the new urban plan is being prepared through the preliminary involvement of the districts.

Other than the reference operational and administrative framework, the players that implemented the process resulting in the realisation of the ETM are worth mentioning. In addition to the foreseeable roles played by various participants (universities, cultural, not-for-profit organisations, etc.), the confederation of craftsmen stood out for their interest in the behaviours and rules to follow in order to not yield to the Mafia's pressure.

Nonetheless, the limits of the ETM held in Brancaccio and in district no. 2 emerged clearly as well. They can be summarised as follows: (high) *cost*, (obsolete) *technology*, *decision makers* (who neglected the final document and tried to exploit it *for political purposes*) and the difficulty in *tackling problems and design solutions for urban space*.

With respect to these critical issues, a few lessons emerged that may contribute to solving the above-mentioned problems: (1) costs can be reduced thanks to further technological and organisational developments, e.g. the TLLS is using the participants' palmtop computers and tablets (Ciaffi et al. 2015); (2) the decision makers must directly attend the training sessions, and they must take part in the participatory events. In this way participants do not feel they have been exploited uselessly and that somebody is trying to exploit their consent; (3) for the participatory processes to be effective, they must be precise and focused on the problems they want to tackle; it takes time to build consistent visions; the risk of *participating with the Mafia* cannot be ruled out unless the promoters in the district are fully aware of their responsibilities; (4) IT experts or technological marketing experts are not enough, as networks of cross-domain skills are necessary; (5) when discussing the quality of urban spaces, technology can be used to obtain suitable external project contributions.

The new administration happened to manage time-consuming red tape that had never before experienced sudden changes and innovations to rules, goals and ways to access the allocated resources in such a short time.

Today the ETMs are indeed the only occasion during which the administration listened to the city in an innovative (for Palermo), organised and transparent way, *also because of smart technologies.*

Although the general goals of participation implementation in Palermo are unassailable, regardless of the majority in the municipal council, various observers pointed out a number of critical issues in the process started by the administration with the cycle of five ETMs. The major problem affecting the effectiveness of the ETMs in Palermo is that they do not stem from the joint action of multiple councils, but rather from the council for participation and administrative decentralisation only.

Some confusion is felt between the participatory instrument and the real context of governing both physical and non-physical transformations at urban, district and neighbourhood level. However, the critical issues are mitigated by an incontrovertible fact. The subjects addressed, the ways in which the assembly definitely changes the initial assumptions of the analyses (detailed in the guides for participants prepared before the assemblies) and, most of all, the contents (in terms of assessment and definition of the guidelines for the administration's policies) are original, relevant and viable from a technical and administrative point of view and mainly focused on the protection and enhancement of existing social capital and not on a vague declamation of impossible utopias.

The local press has harshly criticised the weakness of the ETMs rooted in one single council only, blaming the municipality for not listening to and not involving all the initiatives of participation and active citizen movements in the city.

A no less important aspect about the services offered by ETMs is the training of the officers-trainers in the municipal administration. A fruitful dialogue has been established with them, resulting in the involvement of officers who appreciate this new method of proactive dialogue with citizens as an additional institutional skill. The use of a new know-how on non-ordinary instruments by important human resources of the municipal administration turned out to be of paramount importance.

During the ETMs held in Palermo, the citizens provided politically oriented inputs on urban transformations in a focused, selective, orderly, written, transparent and rapid way. To obtain results characterised by a minimum level of consistency, it is necessary to apply the process and relational approach of HSC to the ETM procedure, as was successfully experimented during the Parterre project. Given the established lack of public financial resources, a long-lasting preparation/training of the human resources who will become the players of the assembly is necessary to *boost* discussion. The ETMs held by the TLLS are the first step of a process intended to involve the municipality in the Digital Agenda.[6]

[6]In this respect, see the European Commission indications on the Digital Agenda: (1) eGovernment Action Plan (2011–2015): *Harnessing ICT to promote smart, sustainable and innovative Government*; (2) Malmö Ministerial Declaration on eGovernment and the support offered by ICT in the life of civic institutions; (3) Europe for Citizens Programme (2007–2013). The trend in Europe of the relationship between politics and active citizenship experiences using ICT has been described in a study carried out by the DG Information Society and Media (Summary of the "Study and

The HSC method experimented in Palermo by the TLLS is likely to ensure a minimum level of transparency in the conflictual relationship between active citizens and the municipal administration. The concerned stakeholders judged the emergence of problems and knowledge data as the achievement of a mutual advantage.

Exchanging information and comparing points of view in the numberless moments of conflict between the representatives of active citizen movements and municipal administration seem nonetheless extremely important, in the light of the widespread mistrust that prevents citizens and decision makers in the municipality, who often have clashing positions, from moving closer.

For the purpose of this work, the expression "HSC" is used for the social capital mobilisation through the conscious use of technologies (from basic to the most advanced ones), in contexts of interaction between players acting in fields made of co-creative ecosystems. Since 2009 and in many different occasions, the partnership of TLLS subjects has been implementing different approaches – which today can be reintegrated in the HSC style – by promoting various initiatives mainly focused on the Digital Agenda. Of these only the ETMs are strongly intertwined with urban planning. The ETM held in Brancaccio represents one milestone of a long process of identity reconstruction in a local community that, eventually, shares the same vision and a relational network, which did not previously exist and is capable of connecting the socially marginalised groups of the neighbourhood to the whole city, as in a sort of advance. This is, however, only the outset of the process and of the cycle of the five subsequent ETMs concerning the whole city, which must be understood to be the gradual adoption of participation in the technical and administrative action of the municipality, with all the weaknesses, misunderstandings and illusions that institutionalised participation inevitably implies.

The representatives of local and regional institutions that govern and manage urban transformations have been involved during the events. It should not be neglected here that the process has been triggered by the local groups' strength of active citizenship. Nevertheless, the experience of the European project boosted the urge for change and reconnecting urban fabrics from a social point of view, as a momentum for decentralisation and, in the view of the renovated urban morphology, as an enrichment of the existing project visions.

Given the strong crisis that Palermo has been going through for a long time now, some success can be at least pointed out, such as the creation of visions of life quality improvement through a shared method within formalised institutional contexts. This element of success is a small step forward for future planning, not only in Palermo but throughout the whole region. HSC can contribute to changing the view that most citizens share, that the city governing body belongs to a dimension *alien* to the one most citizens live in and share. It may not be a major advance, but surely it is better than the decay and the gradual abandonment of the city planning instruments.

Supply of Services on the Development of eParticipation in the EU", European Commission, DG Information Society and Media, eGovernment unit, 2009; pp. 4–5).

References

Aicardi M, Garramone V (eds) (2011) Democrazia partecipata ed electronic town meeting. Incontri ravvicinati del terzo tipo. Franco Angeli, Milano

Arnstein SR (1969) A ladder of citizen participation. J Am Plan Assoc 35(4):216–224

Bobbio L (2002) Le arene deliberative. Riv Ital Polit Pub 3:5–29

Bobbio L (ed) (2008) Amministrare con i cittadini. Viaggio tra le pratiche di partecipazione in Italia. Esi, Napoli

Braida S (1965) Il castello di Favara. Studi di restauro. Architetti di Sicilia (5–6), Palermo

Brown T (2009) Change by design. How design thinking transforms organizations and inspires innovation. Harper Collins, New York

Ciaffi D, Giambalvo M, Lucido S, Tuttolomondo L (2015) Electronic town meeting a Palermo. Dispositivi tecnologici e limiti della partecipazione. StrumentiRes, Anno VII, n° 1, Febbraio

Concilio G, De Bonis L, Marsh J, Trapani F (2012) Urban smartness: perspectives arising in the Periphéria project. J Knowl Econ 4(2):173–256

De Spuches G (2007) Brancaccio come terreno d'azione. Sguardi geografici su un quartiere delle periferie di Palermo. Arch Stud Urbani Reg 90:183–189

Giambalvo M, Lucido S, Tuttolomondo L (2013) A Sud della partecipazione. L'esperienza dell'eTM di Palermo tra cittadinanza attiva e innovazione sociale. Aggiornamenti Sociali 6(2):145–152

Greenfield A (2013) Against the smart city – the city is here for you to use. Do Projects, New York

Hollands RG (2008) Will the real smart city please stand up? Intelligent, progressive or entrepreneurial? City Anal Urban Trends Cult Theory Policy Action 12(3):303–320

Lukensmeyer CJ, Goldman J, Brigham S (2005) A town meeting for the twenty-first century. In: Gastil J, Levine P (eds) The deliberative democracy handbook: strategies for effective civic engagement in the twenty-first century. Jossey-Bass, San Francisco, pp 154–163

Marsh J (2011) The territorial dimension of innovation and the MedLab project. In: Marsh J, Trapani F (eds) MEDLAB in Sicily: an opportunity for social and territorial innovation. Gulotta, Palermo, pp 39–57

Marsh J, Molinari F, Trapani F (2013) Co-creating urban development: a living lab for community regeneration in the second district of Palermo. In: Murgante B, Misra S, Carlini M, Torre CM, Nguyen H-Q, Taniar D, Apduhan BO, Gervasi O (eds) Computational science, and its applications – ICCSA 2013, 13th international conference, Ho Chi Minh City, Vietnam, June 24–27, 2013, proceedings. Part III. Springer, Heidelberg, pp 294–308

Next (2008) Le città nella città Politiche urbane, disagio e devianza minorile alla periferia di Palermo. Rapporto di ricerca. Available at http://www.nuovenergie.org/materiali/Le%20citta%20nella%20citta.pdf. Retrieved 20 Sept 2014

Prescia R (2012) Il complesso monumentale di Maredolce. Il "sollazzo" normanno alla ricerca di un nuovo paradiso. Kalòs, a. XXIV, n. 3, lug-sett., 18–22

Prescia R, Trapani F (2012) Il posto di Maredolce. Un paradiso a Brancaccio. Strategie per la riqualificazione dell'area industriale di Palermo. Esempi Archit 14:1–15

Trapani F (2011a) Maredolce: un Paradiso a Brancaccio. Dal grande passato ai possibili futuri nella porta orientale di Palermo. In: Montagna C (ed) Maredolce. Studiare il territorio di Maredolce-Brancaccio e valorizzarlo come Distretto culturale e turistico. ABC, Sesto Fiorentino

Trapani F (2011b) Integrated programme of urban regeneration. Urban initiative a paradise in Brancaccio. Public-private partnership for an integrated programme of urban development in the eastern outskirts of Palermo. In: Marsh J, Trapani F (eds) MEDLAB in Sicily: an opportunity for social and territorial innovation. Gulotta, Palermo, pp 163–179

ERRATUM

Human Smart Cities

Rethinking the Interplay between Design and Planning

Grazia Concilio and Francesca Rizzo

G. Concilio, F. Rizzo (eds.), *Human Smart Cities*, Urban and Landscape Perspectives,
DOI 10.1007/978-3-319-33024-2

DOI 10.1007/978-3-319-33024-2_16

Volume number, volume 19 included in the PDF.

The updated original online version for this book can be found at
DOI 10.1007/978-3-319-33024-2

G. Concilio, F. Rizzo (eds.), *Human Smart Cities*, Urban and Landscape Perspectives 19,
DOI 10.1007/978-3-319-33024-2_16

Afterword: The Changing Face of Urban Planning: Towards Collaborative and Creative Cities

Charles Landry

Elastic Planning

The world of cities is changing and with it how we plan for them. Every shift in the means of economic wealth creation creates a new social order, a new type of city, new ways of learning and things to learn and new settings where learning takes places. It requires different cultural capabilities.

To plan an industrial city is different from planning a city where knowledge intensity is key or a city where you want to encourage individuals or organisations to be imaginative and to feel commitment to your place.

In the first case, people are seen simply as units of mechanical production; in other cases, people are seen as the key ideas and thus as wealth generators: this in turn is recasting the planning paradigm.

Planning over time has had to have an elastic framework to reflect emerging needs. Now a new form of planning is necessary, because it responds to a novel set of demands. I call this creative city making, and its operating dynamics, its logic and the priorities differ from traditional spatial planning.

In any planning for cities, some basic issues remain consistent. Clearly, traditional planning with its focus on land uses, how space is organised or where differing types of settlement should occur continues as do planning decisions about infrastructures such as roads, utilities like electricity or sewage. For instance, Kamez, a suburb of Tirana in Albania, mushroomed as an informal settlement with no rules over the last 25 years from a greenfield site to a place of 300,000 people. Clearly a more formal physical planning structure is needed to be put in place, such as road infrastructures or utilities as well as how building permits would be agreed. So we cannot decry this form of planning.

Charles Landry advises cities on their future. He has written several books, most recently *The Fragile City and the Risk Nexus*, *The Origins and Futures of the Creative City* and *The Sensory Landscape of Cities*. For more details, see www.charleslandry.com

G. Concilio, F. Rizzo (eds.), *Human Smart Cities*, Urban and Landscape Perspectives 19,
DOI 10.1007/978-3-319-33024-2

These are, though, hardware concerns and this has largely been the way planning has been seen, and those that have determined the look and feel of places have had this background and mindset. I call this the 'urban engineering paradigm of city making'. This focus has shaped the discipline of planning and the skills applied. These were based largely on engineering, architecture and surveying. These groups have a certain mindset that is both positive and negative. The culture of engineering is logical, rational and technologically adept; it learns by doing, and it tends to advance step by step and through trial and error. It is hardware focused. It gets things done. There is a weakness in that this mindset can become narrow, unimaginative and inflexible and forget the software aspect. The cult of the architect is another aspect that can help and hinder how planning and the city evolve. Architects as manipulators and creators of three-dimensional spaces often feel they are the central actors and shapers in making a city. Yet often you ask yourself the question 'do they understand or even like people'. This can lead them to underestimate the contributions other disciplines make, such as those who understand social dynamics and how culture works or even disciplines like anthropology.

Planning has moved along a trajectory. Initially, there was 'the planner', a professional discipline that has existed for just over 100 years, and over time subdisciplines like the spatial, transport or environmental planner evolved. Between them they developed techniques from building codes to infrastructure standards to guide the process of city life. Urban planning would say of itself that its goal is to improve the built, economic and social environments of communities so being the geographical expression of a society's choices. It further expanded and matured as a consequence in three significant ways: community involvement, urban design and place creation and finally sustainability.

The first emerged in the early 1970s based on ideas of people like Saul Alinsky highlighting the need for planning to involve those affected by planning decisions. As noted: 'if you want to know how the shoe fits, ask the person who is wearing it, not the one who made it'. It is increasingly recognised that engagement helps people get the surroundings they want as well as ensuring communities become safer, stronger, wealthier and more sustaining.

The concern with community highlights the planning of social infrastructure, which is the interdependent mix of facilities, programmes, projects, services and networks, in education, health, housing, culture and social affairs that aim to improve quality of life for a community. It moves from only hard infrastructure such as community facilities to include issues of community capital, social inclusion and relationships.

'Urban design', a discipline that is young, perhaps 35 years old, added another dimension in seeking to shape the physical setting for life in cities. With a focus on the art of making places, it has an integrative perspective involving the design of buildings, groups of buildings, spaces and landscapes. This involves understanding the possibilities of blending planning guidelines with how developers make their sums add up and what makes places special. There had been a concern that planning was good at dividing space and not creating place. The notion of urban design has broadened as people seek to involve other disciplines, so 'place making' has become a fashionable term. It implies bringing together all those who can contribute to make a place and that includes those who understand the software of a city.

A space becomes a place when it is imbued with meaning and significance. Place making is an approach to planning, designing and managing public and private space that seeks out the distinctive and special by listening to those who use it, and in this process a vision or story of place is created. You can even make a motorway a place. Richard Reid, a designer, persuaded the New Zealand transport authority to rethink the route of motorway into Auckland, which was to cut off the edge of a volcano sacred to Maoris. His sweeping redesign and landscaping circumnavigated the volcano, and now the interface between the volcano, the motorway and the cycleway as the divider is beautiful and much loved by motorists, cyclists and viewers alike. The resistance to do this by the transport engineers and their governing body was long, but even they now agree the improvement is vast.

Finally, the rise of the sustainability agenda with growing force from the 1970s onwards has reshaped planning goals and the way plans are judged. The concept is now embedded as a matter of course into any urban plan.

The idea of planning with a culturally literate mindset, important as it is, has not yet embedded itself. Nor has a form of planning that recognises the power of creativity and imagination. This brings us to creative city making and we explore its assumptions, aims and challenges in more detail.

The Creative City Making Paradigm

The art of creative city making involves thinking through a number of factors, they include:

- An understanding that planning cities is perhaps the most important human endeavour as it deals with the conditions for people and groups to live together in relative harmony. The skills and insights needed for this combine the physical, cultural, social, psychological, technological and economic domains. Planning education, unfortunately, is still largely focused on the physical aspects and architecture. This needs to change.
- The creative city making concept seeks to be a shift in paradigm. It highlights a different organising map for thinking about cities. It seeks to describe a new wider world of thought and world view of the discipline of planning and urban development. It is an integrated approach which incorporates the specific subsector perspectives such as understanding social dynamics, cultural needs and how the built fabric of a city is put together. In doing this it hopes to create new insights.
- The world of cities is changing dramatically and with it notions of what makes great places and cities as well as ideas of who should be responsible for planning them. There is an older and newer thinking about cities in relation to planning

and design, organisation and management, the use of resources, the role of culture and how cities are marketed. The new planning paradigm should supersede the previously dominant model of urban development focused on the hardware of the city – the 'culture of engineering' approach.

- Successful city making understands urbanity. This is the art and science of how the hardware and the software of a city blend together and how a good physical fabric can assist in generating a positive multilayered experience. The totality of these experiences result in the urban culture of a place. New priorities are required to enhance urbanity involving new competences, a new understanding of what urban assets are and new ways of working to maximise potential.
- A city is not a set of lifeless buildings and roads, but primarily the activities of people who bring it to life. This means only holistic thinkers, and these are rare, and thus perhaps instead integrated teams comprising those with knowledge of people, social dynamics and culture need to work with those who understand transport, engineering, physical planning, recreation and housing issues. The agendas, strategies and plans need to be jointly conceived, planned and implemented. Crucially, the depth, interest and vitality of a city's urban culture create both better communities and ultimately a new source of competitiveness for cities.
- The attributes and attitudes that helped make some cities partly successful in the past may hinder their potential to become great liveable cities in the future.
- The cities we have mostly disappoint. More of them are ugly rather than beautiful, even though we can argue about 'what is beautiful' or satisfactory from a sensory perspective. They tend to look the same with their large tower blocks often sprawling endlessly into the far horizon. There is too much traffic and noise. A few places delight, yet how do we get more of the cities we want? Thinking deeply first about peoples' culture and then their needs and desires and using our imagination to understand these viscerally is a start. A next helpful step is to respond inventively and with creative insight.
- All through history, the city has been the hub for transactions and exchange of ideas, of knowledge, of trade and of services. The city was always the place where mixing and interacting has happened in spite of separations between classes, groups and the rich and poor or the powerful and disenfranchised. The challenge is to foster these exchanges and to seek to bridge the gaps.
- The city has always been a source of problems as well as a laboratory for finding inventive solutions to any problems it creates. Now the special focus should be on creating innovations to heal the environmental distress in cities, to find ways that the diversity of people in cities can coexist in better harmony and to encourage a 360 % holistic perspective to ensure the complexity of the city is fully understood.
- The city was always a place of inventiveness. The difference between the past and today is that cities now are self-consciously trying to encourage and plan the preconditions for creativity. The central and vital characteristic to make this happen is to foster a more open mindset, management style and organisational structure for inventive ideas and projects to flourish. This allows a city to respond

to changing circumstances, to become adaptable and resilient. This requires more 'creative bureaucracies' that are adaptive, open and create the conditions for people to think, plan and act with imagination.

- Creativity is context driven. What was 'creative' in the twentieth century will be different from what is creative in the twenty-first. The priorities 100 years ago might have been public health. Today it might be dealing with wicked problems such as making the most of diversity or climate change.
- What is regarded as creative in one culture and in one circumstance will differ. However, there are principles of creativity that cut across all cultures and time. These include a combination of childlike freshness and deep experience or the ability to connect the seemingly disconnected and to see patterns across different things.
- Creativity on its own is not enough. It only gains true importance when an idea is turned into an applied innovation and reality. A balance of skills is needed for creative city making. However the default position for all those involved in city making should be a willingness to be open and to question assumptions when necessary.
- What is an innovation in one context may be ordinary in another. Turning an old industrial building into a creative economy incubator may be a common practice in Europe but a novelty in India. Or developing a city vision through a public private growth alliance may happen frequently in North America but more rarely elsewhere.
- There are ways of measuring the level and quality of innovations in a city. Some may be paradigmatic shifts with stronger impacts like embedding ecological thinking deeply; some innovations like culture as a driver for economic growth were once a new invention, but now are good practices that most places follow. Over time, a city needs a combination of all these such as paradigmatic shifts, forward-looking innovations and good practices that have been tried and tested elsewhere.
- The innovation journey has a trajectory starting with encouraging people in your city to be curious. If they are curious, they can become imaginative. With imagination it is possible to conceive and reconceive things and to envision possible futures. Then it is possible to be creative and to have ideas. These in turn need to go through the reality checker and to be tested. Only some of the creative ideas will survive and become an invention. Once widely applied, they then become an innovation.
- Today, the central driver and engine for growth of a city is its ability to keep and to attract highly skilled, talented and imaginative people to a city. This is an investment in future prosperity. Crucially such people, who have choices about where to live, increasingly choose the city before the company or job. This means a city needs to rethink its look, feel, appeal and atmosphere to entice these talented and potentially creative people to stay there or be attracted to it.
- Liveability which is the ability to create good facilities and cultural distinctiveness moves centre stage in addressing urban development. The agendas of knowledge nomads and 'ordinary people' align on liveability issues and

the desire for distinctiveness. The wish to be exploratory, which is important pioneering people, is less an issue for most citizens.

- A creative city recognises that an imaginative person can be a scientist, artist, business person, social activist, a bureaucrat, an urban planner or a politician. It is their personal attributes that make them creative rather than what they do. However, creativity is legitimised in some spheres more than others, for instance, the arts of cultural industry sectors like design or new media. This can be a problem even though the creative industries have a special role in helping city development.
- Today the creative economy sectors like design, new media, music, film or the arts as well as the associated cultural infrastructure like museums or galleries can play a central role in fostering the creative urban agenda. They have many impacts including being large job creators, they create positive images for a region, and they are a symbolic factor and are thus a location factor as well as provide enjoyment.
- Creativity is a powerful, soft resource as important as coal or steel or the ability to produce manufactured goods was in a former era. It is a set of attributes that can help create distinctiveness and value in a more knowledge-intensive economy. It can add value to products, services, processes and techniques of doing things as well as how the city itself is shaped and develops.
- Creativity is now seen as a new currency the equivalent to having finance capital. It is equally important. There are also other forms of capital that together can make better cities, including human, social, heritage, knowledge and leadership capital.
- Being creative is a way of thinking and openness is its key attribute. Other characteristics include curiosity; a questioning attitude; the ability to stand back, listen and reassess; and the courage not to take a given credo, practice or theory for granted and to dare to think outside of the box, the gift of seeing relevance and connections between apparently different things. This applies to individuals, organisations and the city itself. Creativity becomes a flexible, multifaceted resource, and its qualities shape peoples' mindset and how they operate as well as the culture of the city. Not every place is equally creative, but everywhere can become more creative than it already is.
- By thinking imaginatively, many new resources and possibilities are uncovered. It is the twenty-first-century version of natural resources. We have moved from a world where natural advantages determined potential to one where harnessing and mobilising creative advantage more effectively than others determines a city's success.
- Imaginative thinking has wide significance cutting across all forms of knowledge including science, technology, and the arts with applications in the social, political and economic domains.
- By creating the conditions for people to think, plan and act with imagination, a creative milieu is established. This is both a physical setting and a set of attitudes that shape the way the city operates. This gives the city its personality. Within this

specific sector may be creative such as a specialism like shoes, machinery, the arts or information technology. However, this alone is insufficient to be deemed a creative city.

- A truly creative place has a 'creative ecology', where all the interconnected systems that make a city function are open to being reassessed when necessary. This might involve those concerned with transport systems, those dealing with social affairs, those developing new business ideas or even managing the city itself. This means being willing to review the tried and tested and to be able to assess well when to open out possibilities and when to close in and be focused.
- Creative city making is concerned with both the software and hardware of the city as well as the 'orgware' of a city.
- The software of the city focuses on its culture, which includes its formal and informal learning systems; the way people across organisations can network, interact and meet; and the atmosphere generated within the city through activities and facilities.
- The city is mostly opaque and obscure. It is difficult to see the city in its totality both from within and how it expresses itself externally. We mostly see its external appearance. You can rarely detect the creative energy often hidden behind walls. Therefore many things stand as a proxy or substitute indicating its creativity. This might be its cultural vitality, its events, gastronomy and cafes or unusual buildings, the streetscape or facilities.
- A city should not simplistically claim it is a creative city. Perhaps it might call itself an 'emerging creative city' as a sign of its aspirations, but preferably it should be others who call your city creative. Here there are elements of push and pull. You push through your intentions, and by creative achievements a virtuous cycle ensues where you are pulled by the external recognition you get.
- A city does not reach an end point when it is then finally creative. The creative city concept is dynamic not static, it is more a continuous process of being alert to opportunities than a detailed plan. The truly creative place knows about timing and balance. It is alert.
- Urban creativity needs a purpose and an aim, which is to give back to its community and even the world and to help a city region to become more resilient, to future-proof itself, to become more prosperous and to enhance its well-being.
- A creative place should have an ethical framework and moral compass that guides its imaginative energies and actions. Today a deep concern to be ecological sustainability is one such element. Another is to develop a humane environment where peoples' needs are primary. A third is to develop a rich multilayered experience for those living in a city. In all of these domains, a vast number of new innovations are required.
- A creative city is not necessarily a comfortable place. Inevitably there will be some tension between the old and the new and between holding onto tradition and fostering innovation. Yet, in an open-minded place which fosters a culture of discussion and debate, this tension can be productive. It can make the city alive and vital.

- There is a deeper impulse that drives our wish to change, adapt, seek improvements or innovate. It is our need to survive and our innate sense of playfulness as human beings. It is this childlike freshness that on occasion we need to recapture.
- We can increasingly assess how creative a city is and the key factors are how a city:
  - Nurtures and identifies its creative potential and reinforces its cultural distinctiveness in order to generate more innovations and so make the city more resilient
  - Enables and supports this creative capacity so that opportunities and prospects are maximised
  - Exploits and harnesses its expertise, talents and aspirations
  - Lives and expresses this in the city through its urban design and related matters
- This can be evaluated through ten domains, which are:
  1. Openness, trust, accessibility and participation
  2. Talent development and the learning landscape
  3. The political and public framework
  4. Strategic leadership, agility and vision
  5. Professionalism and effectiveness
  6. Entrepreneurship, exploration and innovation
  7. Communication, connectivity and networking
  8. Distinctiveness, diversity, vitality and expression
  9. The place and place making
  10. Liveability and well-being

'The City 1.0', 'The City 2.0' and 'The City 3.0'

In order to understand the trajectory towards a creative city better and the complexities of urban transformation, we can describe a simple method. We can call the historic city we have inherited from the past 'the City 0.0', and there are many variations on this. Then, there follows a sequence of 'the City 1.0', 'the City 2.0' and 'the City 3.0'. Most cities need to move decisively from a City 1.0 to a City 3.0 which is essentially the more creative city. Below is a thumbnail sketch of some of their features.

The City 1.0

We can portray 'the City 1.0' in a stereotypical way as follows: The main symbol of this urban type is the large factory and mass production; the mental model is the city as a machine; the management and organisational style is hierarchical and top-down; structures are siloed, vertical with strong departments, and there is little if any partnership; the method of acquiring knowledge is by rote learning and repetition; there is a low tolerance of failure; functions, such as working, living and leisure, are separated; there is little understanding of aesthetics. There is a parallel planning version of 1.0 which focuses largely on land uses; comprehensive development is the preferred modus operandi; and participation is low and not encouraged. Transport 1.0 is largely focused on making the city suitable for the car and pedestrians seem less important. This results in ugly road infrastructures. Culture 1.0 concentrates mainly on traditional forms; cultural institutions dominate; it is reliant on patronage either by wealthy individuals or by the public sector; audiences are quite narrow with elites being the main participants, although folk events are widely popular; culture is seen as detached from commerce.

Overall this is the rational, ordered, technically focused and segregated city. It is the hardware-focused 'urban engineering paradigm' for city making. It reflects a mental attitude and approach to life. It had its highpoints from the 1960s to 1980s. Unfortunately, residues of this approach still exist both in terms of how people go about their business and in terms of the institutions and physical fabric that are still built today. The latter is essentially soulless, rather ugly and lacking any inspiration. These approaches may have been very productive, efficient and relevant to their time, but not anymore.

The City 2.0

'The City 2.0' by contrast has other priorities and evolves from the 1990s onwards. Its industrial emblem is the science park and high-tech industry; its management ethos has flatter structures; partnership working rises in importance as does collaborative working; learning systems open out. There is greater awareness of the need to integrate disciplines. The mental model sees issues as more connected, and this urban form is more aware of how the software and hardware of the city interact. Urban design becomes a higher priority. It begins to focus on the emotional feel of the city and its atmosphere.

There is also an attempt to make the city more spectacular by using new bizarre architectural forms produced by a roving band of nomadic starchitects. Gleaming glass towers proliferate, bold shapes break out of traditional patterns of the square box; skyscrapers explode onto the landscape, some with good public spaces. Vast retailing, entertainment or cultural centres try to bewitch, enchant and seduce you; citizens become more like customers and consumers.

Yet there is also a move to reflect human need and human scale. How people interact rises up the agenda. The city becomes a canvas and stage for activities. Planning 2.0 is more consultative. It sees the city in a more rounded way by linking the physical, the social and the economic, and the notion of transport 2.0 becomes more about mobility and connectivity. The city is less car dominated, and walkability and pedestrian-friendly street design with buildings close to the street become a priority, as do tree-lined streets or boulevards or street parking and hidden parking lots. This 2.0 city seeks to reinsert mixed use and diversity of shops, offices, apartments and homes. It encourages too a diversity of people – of ages, income levels, cultures and races.

Respect for ecology and the value of natural systems rise as do the use of eco-friendly technologies and energy efficiency. More local production is in evidence. There is more emphasis on distinctiveness, aesthetics, human comfort, and creating a sense of place.

Culture 2.0 shifts focus. There is a greater awareness of the power of creative economy sectors and the link between the arts and their role in the broader economy; culture becomes a competitive tool, and it is used to encourage urban regeneration and revitalisation; this increases the popularity of museums and galleries in the quest to change the city's image; activating street life and promoting festivals become part of the cultural repertoire. At the same time, community-driven arts projects proliferate as part of a growing movement of engagement and inclusion.

The City 3.0

'The City 3.0' goes one step further; it takes on the virtues of City 2.0 and is based on harnessing the collective imagination and intelligence of citizens in making, shaping and cocreating their city. It can be called 'soft urbanism' as it takes into account the full sensory experience of the city. In making the city, it considers the emotional impact of how people experience the built fabric and thus is strongly concerned with the public realm, human scale and aesthetics. It understands that blandness and ugliness weaken the city. Its mental model is to see the city as an organism. It is an adaptive city that through its flexibility in operating itself has more chances to become resilient and to future-proof itself. Organisationally, it is more flexible; horizontal and cross-sector working become the norm. There is the recognition that in order to succeed we must sometimes fail. Thus there is a greater tolerance of risk.

Learning and self-development are crucial to the City 3.0. In the City 1.0, knowledge institutions remained factories to drill in knowledge rather than communities of enquiry; they taught specific things rather than acquiring higher-order skills such as learning how to learn, to create, to discover, to innovate, to problem-solve and to self-assess. These are all attributes that artists are good at. This is more likely to trigger and activate wider ranges of intelligences. This fosters the adaptability to allow the transfer of knowledge between different contexts and how to understand the essence of arguments rather than recall out of context facts. Only then can talent be sufficiently unleashed, explored and harnessed.

The City 3.0 too recognises that encouraging entrepreneurship is key to make the city of the future work. Thus, in Economy 3.0, creativity and innovation capacity rise in importance, and the system fosters a start-up culture. Open innovation systems often drive development processes and there is collaborative competition. Microbusinesses and SMEs have far greater importance and the key players are very tech-savvy. These companies have greater impact when they can connect with the mainstream industry to mutual advantage. This urban form is concerned with creating cultural and physical environments which provide the conditions for people to be creative. Thus its industrial emblem is the creative zone or creative quarter.

'Third places' become important, which are places neither at home or an office where it is possible to work on the move. This is part of the 'here and there' and 'anywhere and anytime' phenomena, which is a characteristic of our age. A creative place can be a room, a building, a street or a neighbourhood; yet a creative quarter implies more than one structure. Typically they are anchored around one of the several hundred-old warehouses, breweries, train or bus stations or textile factories that have been rejuvenated the world over. They resonate since they exude memory and physically their spaces are large, adaptable and flexible. This is key since part of this world is a pop-up culture where activities appear overnight and then disappear. Things are less solid and permanent.

Planning 3.0 moves away from a strict land-use focus and is more integrative as it brings together economic, cultural, physical and social concerns. Mixed use is crucial to its planning ethos. It works in partnership and finds interesting methods of participation. It recognises that planning is increasingly concerned with mediating differences between complex issues such as fostering urban growth whilst containing the downsides of gentrification. Citizen participation in decision-making is encouraged, and it takes a holistic approach to identifying opportunities and to solving problems. This ranges from rethinking how policy is made to developing an appropriate regulations and incentives regime that help fulfil aims like becoming a green city or 'cradle to cradle' thinking. Indeed being eco-conscious is part of a new common sense. Equally the idea of being intercultural is vital. This City 3.0 recognises that talent attraction is as important as talent retention. Thus immigration laws are adapted to attract the best from the world.

This City 3.0 uses the available technologies to create smart applications. These are interoperable, immersive self-regulating and interactive devices that tell us how our city is going in real time. These help visualise and track the city in motion. The aim is to use the technical capacities to create a smart economy, a smart mobility and a smart living environment. Making this happen requires smart grids and sensors, open participatory and open data platforms and apps for city services. These help monitor aims like being sustainable. It seeks to have a complete and integrated view of city systems such as energy, transport, health and employment by analysing, gathering citizen feedback and leveraging information across all city agencies and departments to make better decisions. The aim is to anticipate problems, such as traffic bottlenecks or excess energy use, in order to minimise the impact of disruptions to city services and operations. Transport 3.0 moves from a sole mobility focus to thinking about seamless connectivity. This is only possible with smart and rethought governance, where it is necessary to coordinate cross-departmental and cross-agency resources to respond to issues rapidly and in an integrated way.

Culture 3.0 increasingly sees people make their own culture. They are less passive consumers and challenge themselves to enhance their own expressive capacities; they often remix existing work and playfully re-create. They even delve into the source code which in turn enhances their curiosity. Culture is performed in more unusual settings – the street, a local café or a pop-up venue.

These overall trends within the Cities 1.0, 2.0 and 3.0 clearly overlap. Many still display a 1.0 mindset in a world that increasingly operates at 3.0. Planning still has older features as do some working in transport or related disciplines. The cultural institutions of 1.0 coexist with those people who live a 3.0 cultural lifestyle, and thus they need to adapt.

Misalignment and Disconnection

The major fault line usually in cities is the misalignment between an evolving 3.0 world and its economy, culture and social dynamics and the existing operating system that still has several 1.0 features. This creates tensions and misunderstanding and this disconnection needs to be overcome. There is a large grouping in most cities not merely defined by age that can operate globally, that is widely connected and networked, and that understands the new business models driven by the Internet where sharing of ideas is more prominent, which thrives in an open innovation environment and often has a portfolio career. To operate well, they require a responsive regulations and incentives regime. For instance, they need flexible office leasing or rental arrangements geared to project focused work rather than being locked into longer-term contracts. Or they need sympathetic banks or subsidy schemes who appreciate the nature of emerging companies and their ways of working.

Comedia, Sheffield, UK Charles Landry

Index

G. Concilio, F. Rizzo (eds.), *Human Smart Cities*, Urban and Landscape Perspectives 19,
DOI 10.1007/978-3-319-33024-2

Kolibri 360 GmbH
Ferdinand-Jühlke-Straße 7
99095 Erfurt, Deutschland
produktsicherheit@kolibri360.de